C000042260

kullah

ssain

Miah

Homestead plant biod

ladesh

S. M. Atikullah
A. B. M Enayet Hossain
Md. Giashuddin Miah

Homestead plant biodiversity in the coastal region of Bangladesh

LAP LAMBERT Academic Publishing

Publisher:
LAP LAMBERT Academic Publishing
is a trademark of
International Book Market Service Ltd., member of OmniScriptum Publishing Group
17 Meldrum Street, Beau Bassin 71504, Mauritius

Printed at: see last page
ISBN: 978-613-9-45379-5

HOMESTEAD PLANT BIODIVERSITY IN THE SOUTHWESTERN COASTAL REGION OF BANGLADESH

Dr S. M. Atikullah

Acknowledgement

The author finds pleasure to express his indebtedness, deepest sense of gratitude and profound regard to his teacher and major supervisor Professor Dr. A.B.M. Enayet Hossain, Department of Botany, Faculty of Biological Sciences, Jahangirnagar University, Savar for his outstanding designing, guidance and constant encouragement in the long trek of this study and preparation of this dissertation.

The author gratefully expresses his sincere and immense gratitude to his ever enthusiastic teacher and co-supervisor Professor Dr. Md. Giashuddin Miah, Head, Department of Agroforestry and Environment, Bangabandhu Sheikh Mujibur Rahman Agricultural University (BSMRAU), Salna, Gazipur for his continuous encouragement, relentless supervision and guidance throughout the study period.

The author expresses sincere gratitude to Professor Dr. Mahfuzur Rahman, Chairman, Department of Botany for his continuous assistance and Professor Dr. Abul Khair, Professor Dr. Jahed Uddin Mahmud Khan, Professor Dr. Firoza Hossain, Professor Dr. Shaymol Kumar Roy, Professor Dr. Nazmul Alam, Dr. Saleh Ahmmad Khan, Associate Professor, and other teachers of the Department of Botany, Jahangirnagar University, Savar for their positive comments and valuable suggestions during the seminars. He is thankful to all the staff of this Department and general administration section of the university for their cooperation during the study.

He is highly thankful to Mr. Ad Spijkers, Food and Agriculture Organization (FAO) Representative in Bangladesh and to Dr. Ciro Fiorillo, Chief Technical Advisor, Dr. Lalita Bhattacharjee and Mr. B. K. Nandi, Technical Assistant Team Member, and Dr. Nur A. Khondaker, Research Grant Administrator, National Food Policy Capacity Strengthening Program (NFPCSP) of FAO for their support and financial assistance for carrying out extensive field study.

He also extends his vote of thanks to the scientists and authorities of different departments who have provided their assistance by giving secondary data and plant identification and necessary information to complete the study: Soil Resource

Development Institute (SRDI), Bangladesh Meteorological Department, Bangladesh National Herbarium (BNH), SAARC Agriculture Information Centre (SAIC), Bangladesh Water Development Board (BWDB), Bangladesh Bureau of Statistics (BBS), Department of Agricultural Extension (DAE) and Department of Forest and Environment (DoFE). Librarians of Jahangirnagr University and Bangabandhu Sheikh Mujibur Rahman Agricultural University.

He is very much thankful to his wife Sultana Raziatuddin, Daughter Maliha Atik for their inspiration, sacrifice to make the study possible. He also likes to show gratitude to his elder brother Shah Waliullah and sister Fazilatunesa, also to his grand parents M. A. Muttalib Sikder and Aleya Begum for their prayer to Allah for me.

He is thankful to Mr. Jefrie S. Pereira, Retired Executive Director, Caritas, Dr. Quzi Faruque Ahmed, Founder, Proshika, Dr. A. Atiq Rahman, Executive Director, Bangladesh Centre for Advanced Studies (BCAS), Mr. Andrew Jenkins, Team Leader, Integrated Planning for Sustainable Water Management (IPSWAM) and all aunts, uncles, friends, colleagues, and teachers for their encouragement and support for completion of this study.

He pays his deep condolence for the departed souls, who laid down their lives in the memorable cyclone super SIDR of 15 November, 2007 across the coastal zones of Bangladesh. Finally, he offers thanks and gratitude to all the farmers and villagers of Kalisuri, Chaulapara and Nayapara including respondents, teachers, women, and pathfinders of the remotest saline affected areas for support, hospitality and sincere cooperation during the study.

The Author

HOMESTEAD PLANT BIODIVERSITY IN THE SOUTHWESTERN COASTAL REGION OF BANGLADESH

Summary

Homestead is one of the most important natural resource bases of Bangladesh having huge number of diversified plant species. Majority of the people of southwestern region largely depend on homestead production for their survival. Plant species grown and utilized in the homesteads varied in different saline zones of this region. These homestead resources are under tremendous pressure due to recurring natural disasters and human socioeconomic activities. A comprehensive study was undertaken to know the existing homestead plant resources, knowledge base for their management, utilization and conservation in the southwestern region of Bangladesh. The objectives of the study were to: i) document and characterize the existing plant species; ii) find out the relative prevalence, and biodiversity of growing plant species; iii) determine the trend of changing pattern in growing of different plant species; v) identify the management systems and problems faced by the households in their homesteads ; iv) assess the contribution and utilization of trees, and vegetables for the household income and food security; vi) recommend strategies for conservation and sustainable homestead production systems. The present study was carried out in Kalapara and Bauphal Upazillas of Patuakhali district and Amtoli Upazilla of Barguna district considering different salinity levels of these areas. Sampling Unions under these three Upazillas were Latachapali, Karaibaria, and Kalisuri, and one Village under each Union was Nayapara (highly saline), Choulapara (moderately saline) and Kalisuri (less saline). The respondent households were categorized as large, medium, small and landless. Total households of these three villages were 671 of which 36% samples were investigated from July 2006 to March 2008. A pre-tested structured questionnaire, field visits, group discussions and direct measurement of plant diversity were the major tools to achieve the objectives of the study.

A total of 189 different plant species and 62 vegetable-yielding species were found to grow in the tidal saline southwestern coastal homesteads which proved the richness of organismal and ecological diversity. The available plant species were systematically classified which belonged to 76 families of trees and 18 families of vegetable-yielding species representing a wide range of diversity of monocot and dicot plants. The number of trees and vegetable species varied as per salinity level and a total of 29 species were characterized as moderate to strongly saline-tolerant and 19 were found to be tolerant to less saline zones. It is recommended that saline tolerant cultivars and species are to be popularized for improving overall homestead production. On an average, 181 tree species existed per homestead which is encouraging in terms of homestead biodiversity composition. Among the farm categories, landless farmers accumulated a large number of trees for their livelihood and survivability. A total of 11 profitable timber-yielding species, 14 economic fruit-yielding species and 19 multipurpose tree species were found in the study areas. These species were treated as poor man's species for "poverty reduction" and multifarious economic and profitable uses.

Some threatened and rare species were also identified, and these were Urigab/Bangab (*Diospyros malabarica*), Dakur (*Cerbera odollam*), Abeti (*Flagellaria indica*), Bantula (*Hibsicus moschatus*), Bunokakrol (*Momordica chochinchinensis*), Kamrangasheem or Santarkari (*Psophocarpus tetragonolobus*), Cauphal (*Garcinia cowa*), and *Hyptis capitata*. Some homesteads with some uncommon vegetables were found in the "*Rakhain*" community which need to be improved and widely popularized. The most prevalent and top-ranking timber-yielding and fruit-yielding species were *Albizia richardiana*, *Swietenia macrophylla*, *Samanea saman*, *Mangifera indica*, *Cocos nucifera*, and *Phoenix sylvestris*. The most preferred and top-ranking vegetables were *Lagenaria siceraria*, *Carica papaya*, *Lablab purpureus*, and *Cucurbita moschata*. Variation showed in the relative prevalence of major timber-yielding and fruit-yielding plants in various saline zones of the study areas. Prevalence of other species ranked very poor which was an indication of diminishing trend of homestead plant biodiversity in terms of number of different local and indigenous plants. The highest diversity index was found in moderately saline area, followed by strongly saline and less saline areas. The plausible reason of variation may be sought in the biodiversity indices among the areas in which the homesteads of moderately saline areas adopted saline tolerant, moderately saline tolerant as well as less saline tolerant species which could have positive influence to increase the biodiversity indices. There was a positive correlation between plant diversity and farm size in the study sites, but side by side, population of fruit-yielding species showed a negative trend with the farm size in which fruit trees diversity decreased as the farm size increased. It was proved from the study, that landless and marginal farmers were biased on fruit-yielding species for household consumption, food security and sales for regular source of income to meet their daily expenses. The study proved that homesteads of southwestern region are being dominated by timber-yielding species and diminishing trend of fruit-yielding species over time. The changing trend between timber-yielding and fruit-yielding and among other species in each homestead was not in balance. It may create a negative impact in the production, nutrition and income generation as well as livelihood of all the farmers of the study region in the long run.

In total, 37 different types of major species were felled of which timber-yielding and fruit-yielding trees dominated. Majority of trees felled irrespective of farm categories belonged to age limit ranging from 10 years to 20 years. It was reported that landless and small farmers felled their trees in short rotation period due to multifarious socio-economic consequences and emergency needs. However, intensive tree plantation in the homesteads in the last 10 years gained a momentum, while farmers have had planted on an average 92 number of planting materials per homestead. It was found that rural markets were the best source of supplying planting materials for all the farm categories. The findings of the study revealed that farmers usually adopt common indigenous management techniques and practices for application of organic manure, chemical fertilizers, earthing-up, thinning, pruning and fencing. Women play a significant role in homestead production, decision-making process and tree management. The major problems identified in the management and conservation of homestead plant biodiversity were lack of advanced knowledge and technology, pests and diseases attack, improper homestead space planning and utilization, lack of maintenance of embankment and sluice gates, intrusion of saline water, low productivity, recurring natural disasters, and canal siltation etc.

About one third of the total income of the household was used to earn directly from the homestead outputs. This can be looked upon as a "reserve bank" of "food and cash." This income accelerates the reduction of rural poverty. As homestead income is a cross-cutting issue in which contribution of timber selling on total homestead's income was the highest. Food security status of the households between 10 to 20 years ago and present time was compared. The overall food security compared to the past have improved specially the marginal and landless people. It was observed that 89.0% of the respondents have had taken three meals a day and 11.0% two meals a day, indicating a good sign of changing food habit. About 48.0% of the respondents were not involved in food storage and 52.0% of the respondents were involved in subsistence food storage during the rainy season. The common pattern of food habit of having water-soaked rice in the morning and evening is being replaced gradually. However, about 38.0% of the landless and marginal farmers still have had two meals a day is alarming for the poor section of the farmers. The contribution of homestead trees is important to understand its significance in the livelihood support. The major contribution of trees were as supplier of food and fruits followed by cash money, risk coverage and timber and fuel wood etc. The contribution of homestead-based cultivable vegetable was the highest (55.59%), followed by naturally growing vegetables (15.66%). The "life supporting tree and vegetables species" provided enormous opportunity for food security to the people, especially during the crisis period. Combination of farmer's choice and plant diversity perspective shall have to be incorporated to promote quality of planting in the homesteads. Steps should be taken to conserve the following species since their number or population size is decreasing which include Kadamba (*Neolamarckia cadamba*), Royna (*Aphanamixis polystachya*), Jamun (*Syzygium fruticosum*), Devil's tree (*Alstonia scholaris*), Bhola (*Cordia dichotoma*), Custard apple (*Annona reticulata*), Indian dillenia (*Dillenia indica*), Cauphal (*Garcinia cowa*), Chaste tree (*Vitex negundo*), Basil (*Ocimum tenuiflorum*), Seef wood (*Casuarina equisetifoliarum), Indian oak (*Barringtonia acutangula*), etc.

A comprehensive plan has been suggested in view of homestead biodiversity for sustainable production. The components of this plan are: i) massive plantation of diversified species (fast-growing and multipurpose timber-and fruit-yielding species, fruit-yielding species seedlings, commercial vegetable cultivation, local and minor fruit, medicinal and wild species), ii) easy availability of agricultural inputs (organic composting, fertilizers, quality seeds, pesticides, power tiller), iii) application of modern cultivation techniques (better seed preservation techniques, improved varieties, increased production), iv) financial assistance (ready cash, micro-credit, bank facility), v) homestead space utilization (landscape planning, technical assistance, training), vi) improved management of homestead (tree management and improvement), vii) poultry-livestock and fisheries management (technical support, training), viii) disaster management (flood control, disaster preparedness, coping capacity), ix) supplying of safe drinking water (tube well sinking, arsenic and salt-free water, preservation of rain water), x) minimization of salinity (maintenance of sluice gates, embankments and canals, canal re-excavation), xi) government policy support (adoption and monitoring, input support, safety net). A comprehensive space planning, proper scientific management of plants including diversified agricultural and forest cultivars as well as raising critical awareness among the people are very essential to promote and maintain homestead plant biodiversity in the southwestern coastal region of Bangladesh.

TABLE OF CONTENTS

Acknowledgement i
Summary iii
Table of contents vii
List of tables xi
List of figures xii
List of maps xiv
List of appendix xiv
List of boxes and picture plates xv
List of Bengali and vernacular terms xvi
Abbreviations and acronyms xvii
Definition of important terms xviii

	CHAPTER 1: INTRODUCTION	Page
1.1	Rationale	1
1.2	Homestead - a home for plant biodiversity	
1.3	Statement of the problem	
1.4	Objectives of the study	
1.5	Limitations of the study	
1.6	Assumptions of the study	
	CHAPTER 2 : REVIEW OF LITERATURE	9
2.1	Concept and framework of biodiversity	
2.1.1	Genetic diversity	
2.1.2	Ecological diversity	
2.1.3	Organismal/Systematic biodiversity	
2.1.4	Agro-biodiversity	
2.2	Homesteads of Southwestern zone	
2.3	Homestead plant species	
2.4	Utilization of homestead plants	
2.5	Biodiversity degradation/loss of biodiversity	
2.6	Biodiversity conservation	
2.6.1	*In-situ* conservation	
2.6.2	*Ex-situ* conservation	
	CHAPTER 3 : STUDY AREA	16
3.1	Patuakhali and Barguna districts	
3.2	Ecosystems of the study area	
3.3	Characterization of the saline tidal floodplain and coastal marine water zones	
3.3.1	Physiography	
3.3.2	Rivers, canals and wetlands	
3.3.3	Soil	
3.3.4	Salinity level	
3.4	Climate	
3.4.1	Rainfall	

3.4.2	Temperature	
3.4.3	Natural calamities	
3.4.4	Climate change and cyclone Sidr	
3.5	Bio-ecological characters of Southwestern coastal zone	
3.5.1	Land use pattern	
3.6	Land tenuring and leasing system	
3.7	Land leasing system	
3.8	Hydrological constraint	
3.9	Education	

CHAPTER 4 : METHODOLOGY 25

4.1	Area of the study	
4.2	Salinity level of the study sites	
4.3	Site selection	
4.4	Procedure of data collection	
4.5	Questionnaire development	
4.6	Survey	
4.7	Time of data collection	
4.8	Sampling method	
4.8.1	Sample size and sampling	
4.9	Types of data collection	
4.10	Relative prevalence and species diversity indices	
4.10.1	Relative prevalence of species	
4.10.2	Shannon-Wiener species diversity index	
4.10.3	Simpson's species diversity index	
4.10.4	Equitability	
4.10. 5	Correlation among the seven variables	
4.10. 6	Types of species planted	
4.10. 7	Felling trend of trees	
4.10. 8	Household income and expenditure	
4.11	Determination of economic and profitable timber-and fruit-yielding species	
4.12	Determination of multipurpose tree species	
4.13	Discussion meeting	
4.14	Secondary information collection	
4.15	Data analysis	

CHAPTER 5: RESULTS AND DISCUSSION 35

5.1	Household demographic and land profile	
5.1.1	Gender, marital and religion of the respondents	
5.1.2	Age of the respondents	
5.1.3	Education level of the respondents	
5.1.4	Household size of the respondents	
5.1.5	On-farm and Off-farm occupation of the respondents	
5.1.6	Training received by the respondents	
5.1.7	Land holdings of the respondents	
5.1.7.1	Land ownership patterns of the respondents	
5.2	Configuration and space utilization of homestead	
5.2.1	Homestead configuration	
5.2.2	Homestead space utilization	
5.2.3	Pattern of homestead space utilization	
5.3	Species diversity and characterization	
5.3.1	Plant species diversity	

5.3.2	Species composition
5.3.3	Tree density
5.4	Identification of homestead's economic and profitable tree species
5.4.1	Major economic and profitable timber-yielding species
5.4.2	Economic and profitable fruit-yielding species
5.4.3	Multipurpose tree species
5.5	Vegetable species diversity in different seasons and salinity level
5.6	Species characterization
5.6.1	Systematics of the plant species
5.6.2	Systematics of the vegetable species
5.6.3	Characterization of tree species based on salinity level
5.6.4	Characterization of vegetable species based on salinity level
5.6.5	Low cost minor fruits and vegetables
5.6.6	Rare and threatened species
5.6.7	Dominant species in the homesteads of Hindu and Rakhain communities
5.7	Relative prevalence of plant species
5.7.1	Relative prevalence of tree species
5.7.2	Relative prevalence of major species at various saline zones
5.7.3	Tree species reduced to a few individuals in the homesteads of the study areas
5.7.4	Tree species of very low prevalence value
5.7.5	Relative preference of vegetable species
5.8	Diversity indices of different categories of species
5.8.1	Diversity indices of different categories of species in different saline zones
5.8.2	Correlation between age, education, family size and availability of different plant species
5.9	Changing pattern of growing plant species
5.9.1	Changing pattern of growing plant species over time
5.9.2	Tree species per homestead over time
5.10	Plant biodiversity protection and regeneration
5.10.1	Types of species planted during last 10 years
5.10.2	Sources of planting material
5.10.3	Traditional practices in producing seedlings
5.10.4	Types of seeds stored in the homesteads
5.10.5	Indigenous seed preservation/storage techniques
5.11	Felling of trees from the homesteads
5.11.1	Felling trend of trees during the last 10 years
5.11.2	Felling trend of trees based on age groups (maturity) of trees
5.12	Management practices of homestead production system
5.12.1	Management practices
5.12.2	Sorjan-an indigenous technique used for the regeneration of biodiversity
5.13.	Role of gender/women in homestead plant biodiversity management and conservation
5.14.	Problems faced by the farmers in homestead production and management system
5.15	Income and expenditure of the household from different sectors and sub-sectors
5.15.1	Household income

5.15.2   Contribution of homestead income from different sub-sectors
5.15.3   Household expenditure
5.15.4   Contribution of different items of household expenditure and their distribution
5.16   Food security status (food over time, food habit and storage) of the households
5.16.1   Food security over time
5.16.2   Food intake by the households
5.16.3   Stored food in the homesteads
5.17   Relative contribution of the homestead tree and farm products
5.17.1   Contribution of homestead tree product
5.17.2   Contribution of homestead vegetables
5.17.3   Contribution of on-farm and off-farm activities
5.17.4   Important major and minor species for food security
5.17.5   Utilization and status of "life supporting species"
5.17.6   Status of the predominantly grown species in the homesteads
5.17.7   Indigenous knowledge base for homestead plant utilization
5.18   Impact of homestead plant resources on income generation and livelihood support
5.19   Conservation of homestead plant biodiversity
5.19.1   Homestead plant species deterioration compared to the past
5.19.2   Threatened and rare species need to be conserved
5.20   Comprehensive strategies needed to forward homestead biodiversity
5.21   A typical model for sustainable homestead production
          CHAPER 6: CONCLUSIONS AND RECOMMENDATIONS   104
          CONCLUSIONS   104
          RECOMMENDATIONS   108
          CHAPTER 7 : REFERENCES   110
          Picture plates   115
          Appendix   125

LIST OF TABLES

Table

1	Sampling sites (Villages, Unions, Upazillas, and Districts) as per the salinity level of the study areas
2	Distribution of the respondents based on gender, marital and religion at the studied sites of the southwestern coastal region of Bangladesh
3	Household size and sex distribution of the respondents among the different categories of the household
4	On-farm and off-farm occupations of the respondents of the studied southwestern coastal region of Bangladesh
5	Training received by the respondents of the studied southwestern coastal region of Bangladesh
6	Household land holdings of the respondents of the southwestern coastal region of Bangladesh
7	Land ownership patterns of the respondents of the surveyed southwestern coastal region of Bangladesh
8	Homestead construction with different directions at the studied sites of the southwestern coastal region of Bangladesh
9	Level of homestead space utilization at the studied sites of the southwestern coastal region of Bangladesh
10	Homestead plant species diversity in different salinity zones of the studied southwestern coastal areas of Bangladesh
11	Tree composition per homestead on the basis of salinity level in the studied areas of the southwestern coastal region of Bangladesh
12	Tree density on the basis of farm size of different farm categories of the studied southwestern coastal region of Bangladesh
13	Multipurpose tree species uses, weightage and ranking in the study areas of the southwestern coastal region of Bangladesh
14	Vegetable (cultivated and naturally growing) species grown in the studied areas of the southwestern coastal region of Bangladesh
15	Systematic(families, genera and species) of the homesteads' plant species of the studied areas of the southwestern coastal region of Bangladesh
16	Systematic of homestead vegetable species of the studied areas of the southwestern coastal region of Bangladesh
17	Tree species grown in moderate to strongly saline tolerant and less saline areas of the study sites of the southwestern coastal region of Bangladesh
18	Vegetable-yielding species grown in moderate to strongly

saline tolerant and less saline areas of the studied sites of the southwestern coastal region of Bangladesh

19   Low cost minor fruits and vegetable-yielding species grown in the study areas of the southwestern coastal region of Bangladesh

20   Rare and threatened species of the homesteads in the study areas of the southwestern coastal region of Bangladesh

21   Dominant species in the homesteads of Hindu and Rakhain communities of the southwestern coastal region of Bangladesh

22   Species existed in one or more homesteads but reduced to a few individuals in the study areas of the southwestern coastal region of Bangladesh

22.1   Tree species belonging to very low relative prevalence values at the southwestern coastal region of Bangladesh

23   Top ranking most preferred vegetable species grown in the studied southwestern coastal region of Bangladesh

24   Species diversity indices and equitability of different farm categories of the studied areas of the southwestern coastal zone of Bangladesh

25   Species diversity indices of different categories of species in different saline zones of the study areas of the southwestern coastal region of Bangladesh

26   Correlation between age, education, family size and different plant species of the studied areas of the southwestern coastal region of Bangladesh

27   Tree plantation per homestead during last 10 years in the southwestern coastal region of Bangladesh

28   Sources of planting materials used by the respondents of the study areas of the southwestern coastal region of Bangladesh

29   Indigenous seed preservation/storage techniques in the homesteads of the studied southwestern coastal region of Bangladesh

30   Dominant tree species felled in the homesteads during the last 10 years at different farm categories in the study areas of the southwestern coastal region of Bangladesh

31   Felling trend of trees based on age group of trees of the studied areas of the southwestern coastal region of Bangladesh

32   Management practices followed by the respondents in the homestead production system of the studied areas of the southwestern coastal region of Bangladesh

33   Gender/Women role in homestead biodiversity conservation and overall management

34   Major problems faced by the farmers in homestead production and management system of the studied areas of the southwestern coastal region of Bangladesh

35   Household expenditures according to farm categories of the studied southwestern coastal region of Bangladesh

36   Food security of the respondents over time in southwestern

	coastal region of Bangladesh
37	Food intake by the households of the studied areas of the southwestern coastal region of Bangladesh
38	Stored food in the homesteads by the different farm categories of southwestern coastal region of Bangladesh
39	Contribution of homesteads trees for livelihood support of the respondents of the studied areas of the southwestern coastal region of Bangladesh
40	Contribution of homestead's cultivated and naturally growing vegetables in the homesteads of studied areas of the southwestern coastal region of Bangladesh
41	Contribution of On-farm and Off-farm activities of the southwestern coastal region of Bangladesh
42	
	Homestead species dominant with regard to food security of the study areas of the southwestern coastal region of Bangladesh
43	
	Indigenous knowledge base on homestead plant utilization
44	Homesteads plant species deterioration compared to the past in the study areas of the southwestern coastal region of Bangladesh according to respondent opinion
44.1	List of minor non-woody plant species of the homesteads which are diminishing compared to the past in the study areas of the southwestern coastal region of Bangladesh
45	Threatened and rare species need to be conserved in the study areas of the southwestern coastal zone of Bangladesh

LIST OF FIGURES

Figures

1	The concept and level of biodiversity
2	Total rainfall of Patuakhali and Khepupara stations under Patuakhali and Barguna districts during 1983 to 2007
3	Temperature data of Patuakhali and Khepupara under Patuakhali and Barguna districts during 1983 to 2007
4	Centre of the cyclone Sidr which affected the study area
5	Salinity status of various upazillas belonging to Patuakhali and Barguna Districts
6	Age category of the respondents (%)
7	Education levels of the respondents (%) of the studied southwestern coastal region of Bangladesh
8	Pattern of homestead space utilization in the study areas of the southwestern coastal region of Bangladesh
9	Economic and profitable timber-yielding species in less saline to strongly saline areas of the studied southwestern coastal region of

Bangladesh

10	Economic and profitable fruit-yielding species in less saline to strongly saline areas of the studied southwestern coastal region of Bangladesh
11	Relative prevalence of dominant species (timber-yielding, fruit-yielding, medicinal, ornamental and naturally growing) in the studied sites
12	Relative prevalence of major timber-and fruit-yielding species in different saline areas of the studied southwestern coastal zones of Bangladesh
13	Changing pattern of growing of different plant species in the homesteads of the studied areas of the southwestern coastal region of Bangladesh
14	Existing tree species per homestead in the studied areas of the southwestern coastal region of Bangladesh
15	Types of seed collected and stored in the homesteads of the studied areas of the southwestern coastal region of Bangladesh
16	Household income of the respondents from different sources (expressed in %) of the studied southwestern coastal region of Bangladesh
17	Contribution of homestead income from different sub-sectors by all the farm categories of the studied southwestern coastal region of Bangladesh
18	Proportion of different items of household expenditure in the homesteads of the study areas of the southwestern coastal region of Bangladesh

LIST OF MAPS

Map

| 1 | Showing the Southwestern coastal region of Bangladesh |
| 2 | Map showing the selected three upazillas having different types of salinity level of the study sites under the two peripheral districts facing Bay of Bengal |

LIST OF APPENDICES

Appendix

| 1 | Transect of the studied villages (Nayapara, Choulapara and Kalisuri) |
| 2 | Questionnaire of the study |

3	List of homestead plant species identified and characterized in the study areas of the southwestern coastal region of Bangladesh
4	List of homestead vegetable-yielding species identified and characterized at the study areas of the southwestern coastal region of Bangladesh
5	Relative Prevalence (RP) of all tree species in varying saline zones of southwestern coastal zone of Bangladesh
6	Type of tree species planted during the last 10 years at different farm category of the study areas of the southwestern coastal zone of Bangladesh
7	Felling trend of trees in the homesteads during the last 10 years at different farm category in the study areas of the coastal zone of Bangladesh
8	Homesteads plant species deterioration compared to the past in the study areas of the southwestern coastal region of Bangladesh

LIST OF BOXES AND PICTURE PLATES

Box/
Picture

1	Impact of homestead plants on income generation and livelihood support of the study areas of the southwestern coastal region of Bangladesh
2	Comprehensive steps needed to be taken for homestead biodiversity utilization and conservation in Bangladesh
3	A framework of homestead production through eco-friendly utilization and conservation of natural resources
1	Rural homesteads utility and uses
2	Different uses of economic and profitable fruit-yielding species
3	Different economic and profitable species and their uses
4	Different economic and life supporting species
5	Different rare and uncommon species
6	Diversified homestead species of Southwestern region of Bangladesh
7	Seedling and seed raising techniques, fruit diseases and Sidr damage of homestead resources
8	Different 'life supporting species' as alternative food and income
9	Important photographs on income and employment and a Rakhain family
10	Important photographs on FGD, seminar, field survey and elderly persons

List of Bengali and Vernacular Terms

Vernacular name

Choutha	It is a term which means four portions by which the crop yield is distributed in four parts
Boi	The stolon of Arum/Kachu
Veranda	A term is used to identify the front or back wart areas of a house
Bigha	A variable unit of land measurement. A standard bigha is one-third of an acre
Borga	A term is used for shared cropping system
Dola/Duli	A indigenous bamboo-made basket used for grains preservation
Do-tala bari	One type of tin-shade two- storeyed house prepared in these areas
Drum/Kati	Iron-made big container used for preserving seeds
Dhankarali	Land given to the tenant for paddy (dhan) as a pre-commitment (karal-oral contact)
Eksona	Land leased only for one year to the tenant
Ek-tala bari	One type of tin-shade one- storeyed house
Gur	Unrefined, locally made molasses (harvested from *Golpata* plant)
Kacha	House floor made with soil
Kandi	Raised border of homesteads prepared for tree plantation and hedge
Kani	Unit of land measurement
Khana	A household, popularly described as 'Khana'
Khaota	2 years land owner can claim to get free the contact by deducting two years money proportionately (i.e. locally termed as 2 years "khaota")
Khas land	Government land which is distributed among the landless
Moa	Indigenous food prepared by fried mixing with molasses
Mocha	The inflorescence of banana termed as 'mocha' is also used as vegetable, very popular food in the southern region
Mogh	Rakhain community are locally called Mogh in coastal areas
Motki	Earthen made big-sized container prepared for preserving seeds
Nagad lagai	Land sale by the land owner for a season in direct hard cash payment
Patta or Khaikalasi	Land leased for seven years or more is termed as Khaikalasi

Rabi crop	Crops grown in winter season (November to March)
T. Aman	Winter transplanted rice in Bengali. Rice varieties cultivated in monsoon season and harvest in post-monsoon
T. Aus	Autumn transplanted rice in Bengali. Rice varieties cultivated in pre-monsoon season and harvest in monsoon season
Thor	The peduncle of the spadix inflorescence known as 'thor' used as an indigenous vegetable
Te-bhaga	Crops shared into three portions is termed as te-bagha

Abbreviations and Acronyms

AEZ	Agro Ecological Zones
BSMRAU	Bangobandhu Sheikh Mujibur Rahman Agricultural University
BAU	Bangladesh Agricultural University
BRAC	Bangladesh Rural Advancement Committee
BCCSAP	Bangladesh Climate Change Strategy and Action Plan.
BARC	Bangladesh Agricultural Research Council
BFRI	Bangladesh Forest Research Institute
BSS	Bangladesh Bureau of Statistics
BTRI	Bangladesh Tea Research Institute
C	Celsius/Centigrade
CBD	Convention on Biological Diversity
Cm	Centimeter
DAE	Department of Agriculture Extension
DoEF	Department of Environment and Forest
DFID	Department for International Development
dS/m	Decisimen/miter
FAO	Food and Agriculture Organization
FGD	Focus Group Discussion
GOB	Government of Bangladesh
GOs	Government Organizations
HYV	High-Yielding Variety
Ha	Hectare
IPCC	Intergovernmental Panel on Climate Change
IUCN	International Union for Conservation Nature and Natural Resources
JAO	Junior Agriculture Officer
Km/Km^2	Kilometer/Square kilometer
M	Meter
NAP	National Action Plan
NAPA	National Adaptation Program of Action
NEMAP	National Environment Management Action Plan
GB	Grameen Bank
NGOs	Non Government Organizations

NPK	Nitrogen, Phosphorus, Patassium
NP	National Park
BRDB	Bangladesh Rural Development Board
FD	Forest Department
BRAC	Bangladesh Rural Advancement Committee
PRSP	Poverty Reduction Strategy Plan
RP	Relative Prevalence
SAIC	SAARC-Agricultural Information Centre
SAARC	South Asian Association for Regional Cooperation
SPSS	Statistical Package for Social Science
Sidr	Cyclone of 15 November 2007
Sq. Km.	Square Kilometer
SRDI	Soil Resource Development Institute
T.Aman	Transplanted-aman
T-Aus	Transplanted-aus
UNCED	United Nations Conference on Environment and Development
UNEP	United Nations Environment Program

Definition of Important Terms

Biodiversity : According to Piedad Cabascango, Ecuador, "to me biodiversity is all the beings that are related in nature: man, animals and plants, even vegetables, rivers, seas, animals in the jungle and all the beliefs we have kept from our ancestors and from our dreams. Wisdom itself is also a part of biodiversity (DFID, 2002)."

Cultivar: A cultivated variety (genetic strain) of a domesticated crop plant.

Ex-situ conservation: Keeping components of biodiversity alive away from their original habitat or natural environment.

In-situ conservation: The conservation of biodiversity within the evolutionary dynamic ecosystems of the original habitat or natural environment.

Flora: All of the plants found in a given area.

Homestead: The typical homestead in villages is mostly aggregated on raised platforms usually above normal rain water flooding. Almost all villages are rich in groves of tree species. Common homestead plants are the trees and shrubs that give a lush green panoramic view in Bangladesh.

Household: A household means a group of persons normally living together and eating in one-mess. UN (United Nations) defined household as, "Person or persons related or unrelated, living together and taking food from the same kitchen is considered as a household" (BBS, 2001). Popularly, it is described as "Khana".

Less-saline: This area covers part of Patuakhali and Barguna districts by different degree of soil salinity. Soil salinity in the dry season ranges from 2.1 to 8.10 dS/m.

Moderately-saline: Almost the whole area is devoured by degree of soil salinity. Soil salinity in the dry season ranges from 8.1 to 12.0 dS/m.

Strongly-saline: This area is affected by different degrees of soil salinity. Soil salinity in the dry season ranges from 12.1 to 16.0 dS/m. or > 16.0 dS/m.

Multipurpose trees: A species can be multipurpose in one situation. Multipurpose trees are defined as woody perennial that are purposefully grown to provide more than one significant contribution to the production and/or service functions of a

CHAPTER 1:
INTRODUCTION

1.1 Rationale

Biodiversity is one of the foundations for the development of a country. It has become a very common and useful term since the Conference of the United Nations on Environment and Development (UNCED) held in 1992 in Rio de Janerio (the 'Rio Summit'). Biodiversity creates and contains visible and invisible everything to sustain life on earth. According to the Rio Summit, it is the variability among the existing living organisms (plants, animals and microorganisms) from all sources (terrestrial, marine and aquatic) and the ecological complex of which they are a part. It includes diversity within species, between species and ecosystems. It is the complicated mosaic of living organisms that interact with biotic substances and ecological gradients to sustain life at all hierarchical levels (McNeely, 1990).

Biodiversity is the wealth and the primary source of livelihood for both human beings and animals. Plant biodiversity is the most important and invaluable component of total biodiversity. It is considered as a resource and property of the Plant Kingdom. Plant biodiversity, either in wild and domesticated form, is the source of human food, medicine, energy, clothing and shelter. It has a great role in nature to keep it in balance, to produce biomass in the biosphere and to maintain the energy-flow in bio-geochemical cycles. It maintains the soil fertility and keeps the biotic-abiotic relationship for ecosystem sustainability. In fact, it is the base on which the whole living being depends for survival.

The loss or degradation of biodiversity is a global issue and there is hardly any region on earth where loss of biodiversity has not occurred. Among the known 1.5 million species inhabiting the world, one-fourth to one-third is likely to be extinct within the next few decades. Every single species of creation belongs to nature and there exists an ultimate deep-rooted interrelationship and interdependence amongst all forms of lives. The loss of any creature, small or big, affects the ecological balance, which eventually may lead to a catastrophic end. Therefore, conservation of biodiversity is a critical issue to address and to bring into national as well as individual's cognizance locally, regionally and globally.

Bangladesh, though a small country (area wise), is endowed with a rich reserve of plant resources. It is the abode of different plant species. It has been listed of about 3,611 flowering plants species (Ahmed, 2009). The homesteads of southern region of Bangladesh are prepared above the flood level and have huge space for diversified vegetation that is significantly different from other parts of Bangladesh (Razzak, 2001). This rich phytodiversity in Bangladesh is on the verge of rapid decline, because of rapid depauperation of different species through natural processes. Forest land which was the best home of plant diversity, is under severe threat. Although officially forest land is 17% of the total land area of Bangladesh, but actual tree coverage has been depleted to only 6% due to massive human interferences (Miah et al. 1990). Similarly, farm lands where thousands of cultivable crop varieties are grown with association of wild life and feeding habitats historically are on reducing trend because of intensive land use, especially growing of high yielding and hybrid varieties with high input-based technologies. As a result, the country once endowed with thousands of diverse species has already lost some of them or are in the process of extinction. In spite of that some efforts are made to promote tree plantation in many degraded forests and private lands, homesteads and farm land as a source of diversified plant resources.

1. 2 Homestead - a home for plant biodiversity

Homestead plays a very important role in the livelihood, rural employment and income generation of the country. Bangladesh consists of 87363 thousand villages (BSS, 2007), and each village encompasses a few hundreds of homesteads. They constitute the centre of socioeconomic activities and traditional cultural heritage of villages. Homestead is the most important production unit in Bangladesh having about 25.49 millions households with 19.45 millions in the rural areas (BBS, 2006). These homesteads occupied about 0.54 million hectares of land (BBS, 2001) and this figure is increasing at the rate of 5m^2/ha/year (Anam, 1999). The size (average) of the rural homestead is very small (0.02 ha), but varies widely according to region and socio-economic status of the households.

2

Homestead farming or homestead agroforestry is an age-old practice and an integral part of farming system. It is a complex production system where different plant species including crops or vegetables are grown in association with trees in admixture with or without livestock or fish. A good number of plants and vegetables are naturally grown and exist around the homesteads called: **"life supporting species"**. Uncultivated and natural sources of food, fodder, fuel and vegetables from an important resource base for the homestead.

A large number of higher plants have been reported in homesteads and surrounding areas. **Islam and *et al.* (2013) identified 32 fruit yielding and 37 timber tree species in this coastal region.** Latif *et al.* (2001) identified 148 species of indigenous and exotic types in the village forests. Similarly, Basak (2002) identified 105 tree species and 27 herbaceous species (vegetables and spices) across the major four ecological regions of Bangladesh. Life supporting species provide enormous opportunity of food security to householders, especially during the crisis period. Homestead plays an important role in Bangladesh economy and provides 50% cash flow to the rural poor (Ahmed, 1999). Collectively, homestead production system contributes about 70% fruit, 40% vegetable, 70% timber and 90% fire wood and bamboo requirements of Bangladesh (Miah and Ahmed, 2003). In addition, the home gardens (kitchen gardens) are recognized as repositories of non-timber products, such as medicinal and aromatic plants, ornamentals, bamboos, honey, cane and grasses. The land areas for field crops have declined, while average homestead area per farm has increased from 0.08 to 0.09 acres. This indicates an increased opportunity created to some for extending home-based farm and non-farm production system (Mandal, 2003). Although total areas covered by the homesteads are steadily increasing due to building up of new houses to accommodate the increased population, but on the other hand, homestead plant resources are under increasing pressure of exploitation.

1. 3 Statement of the Problem

Homestead is the most important natural resource base in Bangladesh containing a large number of diversified plant species. Some of these plant species are called "life supporting species" because these species in the homesteads play a vital role for the livelihood of the millions of people, especially during food scarcity in the rural areas.

3

Unfortunately, these resources are under tremendous pressure due to various human activities and frequent natural calamities. In many places, there has been a rapid depletion of homestead resources. All these activities are creating negative impact on the homestead production system and food security of rural households. Sustained production system is already severely affected and in some cases degraded. **The Red Data Book (Ara *et al.* 2013) of vascular plants of Bangladesh listed 120 plant species which faces threats in varying degree. The loss or degradation of every single species which deplete biodiversity ultimately affects the ecological balance and food security of a community.**

Species richness or plant diversity varies from place to place, largely influenced by ecological and socio-economic factors. It varies among the homesteads even within similar ecological and socioeconomic groups depending upon individual needs and preferences. Across the country, coastal region has speciality because of its unique climate.

Coastal region, especially the Southwestern part of Bangladesh i. e., Barisal Division comprising of six districts lying across the Bay of Bengal (Map-1) is bio-ecologically ever dynamic and enriched by rivers and tributaries, open water areas of char lands and homestead plant resources. However, recurring threats of cyclone, flood, storm and tidal surges are the major constraints that commonly break down the coastal ecosystems, specially threatening both the flora and fauna of these areas. These consequently are affecting homesteads of southwestern zone of the country. Homestead of this region is medium and larger in size and possesses luxurious vegetation compared to the other regions. Majority of the people of this region greatly depend on their homesteads, because most of the arable lands are low and suffer from salinity problem. However, recurring threats of disasters are the major constraints that commonly break down the coastal ecosystems, specially threatening both the flora and fauna of these areas. As a ready reference devastating super Sidr, 2007 and Aila 2009 damaged growing plants of this coastal region. On the other hand, drought in the southern region seriously affected homestead vegetations. Miah and Bari, 2001 showed that both number and productivity of homestead plantations, particularly sweet water-loving fruit species (*Spondias pinnata, Manilkara zapota, Areca catechu, Phoenix sylvestris etc*) are declining due to increasing soil salinity. On account of human activities and poor management strategies, many plant species of the region are decreasing trend. Unfortunately, there was little study on the existing homestead species resources for management and conservation, especially in the southern coastal region of Bangladesh.

A comprehensive study needs to be conducted to enumerate and assess plant biodiversity for appropriate management of the households in the coastal region in Bangladesh.

Keeping the above facts in mind, a study entitled "Homestead plant biodiversity in the southwestern coastal region of Bangladesh" has been undertaken to collect systematic information on natural resources and to make necessary recommendations for future intervention by the different national and international actors to sustain or enrich the homestead ecosystem as a prime source of livelihood support.

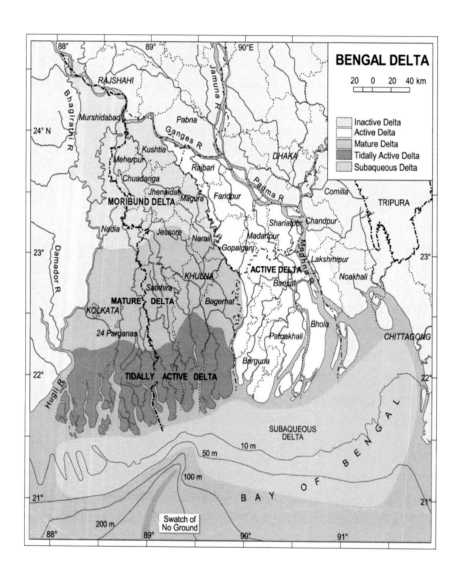

Map-1. Showing the Southwestern coastal region of Bangladesh.

1.4 Objectives of the study

The study was conducted with the following objectives:

i. To document and characterize the existing plant species (herbs, shrubs, climbers and trees) in the homesteads of varying saline zones of the southwestern coastal region of Bangladesh.

ii. To find out the relative prevalence and biodiversity of growing plant species in the homesteads of varying saline zones of the southwestern coastal region.

iii. To determine the trend of changing pattern of growing different plant species in the homesteads over time.

iv. To identify the management systems and the problems faced by the households in cultivating plants in the homesteads.

v. To assess the contributions and utilization of growing trees and vegetables on the household income and food security.

vi. To recommend strategies for conservation of plant and sustainable homestead production system with diversified plant species.

1.5 Limitations of the study

i. The challenge for this study was the natural catastrophe i.e., memorable cyclone Sidr occurred on 15 November, 2007 which seriously affected the study areas, where homestead plant resources were damaged to a great extent.

ii.The study areas were extremely remote which were aggravated by Sidr.

1. 6 Assumptions of the study

The following assumptions were kept in mind at the time of study to achieve the desired goals and objectives:

i. Relevant and desired questions were included in the questionnaire to get the proper responses from the respondents.

ii. The respondents from whom the data were collected were representative of the whole population of the study areas.

iii. The selected areas were the typical ones representing the whole region.

iv. The findings of the study were expected to be useful for sustainable management of homestead biodiversity.

CHAPTER 2: REVIEW OF LITERATURE

This chapter deals with the concept, review of past studies and findings relevant to the present study. In fact, very few research works were conducted so far on homestead plant resources, especially in the southwestern saline zones of Bangladesh. However, a few literature were collected from different sources which were presented below in six major sections. The sections were noted as concept and framework of biodiversity, homesteads of Southern zone, homestead plant species, biodiversity degradation/ loss of biodiversity, and biodiversity conservation.

2.1 Concept and framework of biodiversity

Biodiversity means the variability among living organisms from all sources and the ecological systems. It can be partitioned to make out the status of biodiversity of a country, of an area, or of an ecosystem, of a group of organism, or within a single species (UNCED, 1992). Wisdom (human cultural dimension) itself is also a part of biodiversity (DFID, 2002). According to Heywood and Watson (1995), it is composed of genetic, ecological and systematic diversity. As a new element of human life, biodiversity can be set in a frame (Figure-1) so that species extinction, disappearance of ecological association, or loss of genetic variation that is extant, all can be included.

2.1.1 Genetic diversity

It refers to the variation of genes within the species. This constitutes the distinct population of the same species or genetic variation within population or varieties within a species.

2.1.2 Ecological diversity

Ecosystem diversity could be understood if one studies the communities in various ecological niches within the given ecosystem; each community is associated with definite species complexes. These complexes are related to composition and structure of biodiversity.

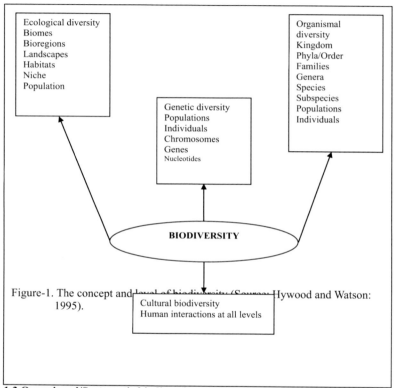

Figure-1. The concept and level of biodiversity (Source: Hywood and Watson: 1995).

2.1.3 Organismal/Systematic biodiversity

Taxonomy is the practice by which the different groups of organisms are described and classified in a hierarchical system and thereby provides a reference system for biology. Scientific classifications are the means for dividing the world organisms into different units about which one can communicate. Systematic biology is the approach by which similarities and differences between these groups are examined and their evolutionary relationships are elucidated. The applied science of systemics include the planning, conservation, prospecting, exploitation, regulation and sustainable use.

2.1.4 Agro-biodiversity

Agro-biodiversity has been fast emerging which deals with the life forms of agro-ecosystems. Diverse agro-ecosystems of Bangladesh are rich and major repository of agricultural genetic pools. Local communities have selected and conserved genetic variation in plants in the various agro-ecological zones and small farmers have played a major role to develop, adopt and preserve these genetic bases. The greatest diversity is present in rice (*Oryza sativa*). There are about six thousand varieties of rice known to have existed in the country (Kumar and Mahendra, 2000).

2.2 Homesteads of Southwestern zone

The homesteads of southwestern region are typically different in which tin-shade houses are built up to two (do-tala) or one (ek-tala) storeyed structures raised from the ground in high and medium high platform for protecting from flood, storm, and tidal surge etc. People usually plan to prepare a homestead usually in an encircled area and raise the edge (Kandi) of a boundary to plant hedge and shade species. The main house encompasses a corridor (Veranda) for the protection of the main house and for multipurpose uses also. Most of the houses are *Kacha* that are very common in these areas.

In the saline system, homesteads are the only place on which majority of the people depend. During natural disasters such as flood and cyclone, homestead is the only one place where people can get shelter and protect themselves. Especially the poor and pro-poor affected during this period, are fully dependent on available plants and vegetables in the homestead. Therefore, plantation and management are the important livelihood activities of many landless and small households. The current research initiative was taken to address the problems of plant biodiversity in the southwestern coastal zone which differs considerably from the non-saline areas of the country's mainland.

2.3 Homestead plant species

Identification is one of the important part of plant taxonomy to document and characterize existing homestead plants. This is most applicable to the southwestern zone of Bangladesh as this region is naturally vulnerable. Alam *et al.* (1996) identified about 183 species belonging to 136 genera and 48 families in ecological zones of the country. Uddin *et al.* (2001) identified 62 useful plant species in the homesteads of saline areas of southern Noakhali district. Among them, there were 30.9% fruit, 29.09% timber, 34.54 % vegetable and 5.45% spice-yielding species. Alam and Masum (2005) also found 142 plant species belonging to 61 families in the offshore island of southern zone of Bangladesh. Kabir (2007) investigated the floristic and structural diversity of 402 homegardens from six regions across southwestern Bangladesh. Each region contained a mean of 293 species out of 67 homegardens. A total of 49,478 individuals (107 per homegarden and 1003 per hectare) of trees and shrubs were counted from 45.2 ha total sampled area. Millat-e-Mustafa and Haruni (2002) identified a total of 162 perennial species from 100 homegardens surveyed in four regions of Bangladesh. The highest number of species was recorded from the deltaic region (91) followed by the plains (79), hilly (73) and the dry land (65) regions, respectively. Side by side, south Asian and other countries like Pakistan, India and Cuba also showed richness in plant biodiversity in their homesteads. Wezel and Bender (2002) investigated the cultivation of different plants in homegardens for self-sufficiency in Cuba, but knowledge about homegardens in Cuba was limited. The plants studied in the homegardens totalling to 101 different plant species were found with an average number of 18-24 species per homegarden. Sheikh *et al.* (2002) reported 153 plant species belonging to 38 families and 113 genera after extensive field research in the Natar Valley of Pakistan. Kumar *et al.* (1994) noted that there was tremendous variability both in number of trees and shrubs present and species diversity of the selected homesteads in different provinces. In total, 127 woody species (girth at breast height more than or equal to 15 cm.) were reported. The mean number of woody taxa found in the homegardens ranged from 11 to 39. Therefore, homestead trees were important to understand its significance in the livelihood system in diverse ways. The variation in number and different kinds of plant species in less saline to strongly saline areas of the present study area can help realizing the way forward to proper utilization and conservation.

2.4 Utilization of homestead plants

Homestead has social and economic importance especially in the life supporting system. Reviewing the utilization and importance of homesteads, Douglas and Hart (1973) stated that the trees constitute an interesting element of homestead as well as nature. Trees provide direct and also indirect benefits to human being and to nature. It has the great potential for feeding men and for regenerating the soil for restoring water system, for controlling floods and drought, for creating more benevolent micro-elements and comfortable and stimulating living condition for humanity. Homestead as an ecological unit consisting of land, pond, house, plants and animals, which are in continuous interactions with the farmer and his family for fulfilling some of the daily household needs. Homestead, the most important source of natural resources, plays an important role in Bangladesh economy and provides nearly 50 percent cash flow to the rural poor (Ahmed,1999). Properly managed homestead can alleviate poverty of rural people by increasing income. Akhter *et al.* (1989) mentioned that the farmers consider the trees as savings and insurance against the risk of crop failure and low yield. Therefore, homegarden production system collectively contributes about 70 percent fruit, 40 percent vegetable, 70 percent timber and 90 percent firewood and bamboo requirements of Bangladesh (Miah and Ahmed, 2003). Therefore, a systematic study would facilitate finding out and explore the role of homestead plants in the southwestern coastal region for addressing food security of the people of this region.

2.5 Biodiversity degradation/loss of biodiversity

There are so many factors responsible for the degradation/loss of plant biodiversity and for which many species are now becoming threatened and many of them have disappeared. IUCN (International Union for Conservation Nature and Natural Resources), Bangladesh has identified major threats of biodiversity to be the loss of habitat, over harvesting of resources, increasing productivity through monoculture and natural calamities etc. The causes of biodiversity degradation in Bangladesh are pollution, lack of awareness, land tenure and user rights issues, institutional capacity, and human population growth etc. The global biodiversity strategy identified six fundamental causes of biodiversity loss and these are: i) Unsustainable high rates of

13

human population growth and natural resource consumption, ii) Steadily narrowing spectrum of traded products from agriculture and forestry, and introduction of exotic species associated with agriculture, forestry and fisheries, iii) Economic systems and policies that fail to value the environment and its resources, iv) Inequity in ownership and access to natural resources, including the benefits from use and conservation of biodiversity, v) Inadequate knowledge and inefficient use of information, vi) Legal and institutional systems that promote unsustainable exploration (WRI/IUCN/UNEP 1992). Climatic change i e. rising sea level, cyclone and drought in the southern region has also affected homestead plant biodiversity. Recent cyclone super Sidr had lost innumerable number of plants and decreased the productivity of homesteads, particularly in less saline to strongly saline areas due to intrusion of saline water in the homestead areas.

2.6 Biodiversity conservation

Conservation is a vital issue for utilization of natural resources for feeding human population and survival for all living entity. The notion of biodiversity is closely associated with that of its conservation, preservation and sustainable use, especially following the publication of the World Conservation Strategy (IUCN, 1980).

Conservation of biological diversity can be achieved in a number of complementary ways of two separate approaches, *in-situ and ex-situ* methods.

2.6.1 *In-situ* conservation

In-situ means the natural, original place or position which includes conservation of plants in their native ecosystem or even in a man made ecosystem, where they occur naturally. The *in-situ* conservation, natural habitats have received high priority in the world conservation strategy programs launched since 1980 (Kumar and Mahendra, 2000). *In-situ* conservation is a much more open and integrated approach to maintain maximum amount of diversity (ecosystem or species). For domesticated species, *in-situ* conservation means the conservation of traditional farming systems with the exclusion of modern varieties. For wild species, *in-situ* conservation is supposed to be the normal way of protection (Heywood and Watson, 1995). Different categories of *in-situ* conservation are National Parks (NP), Sanctuaries, Biosphere Reserves, Reserve Forests etc. There are four national parks in Bangladesh, and these are: i)

14

Himchari NP, Cox's Bazar, ii) Bhawal NP, Gazipur, iii). Madhupur NP, Tangail, and iv) Ramsagar NP, Dinajpur. The area of these four parks are 1,729 ha, 5022 ha, 8436 ha, and 52 ha, respectively.

The Sunderbans is one of the World's Heritage Sites which acts as the largest *in-situ* conservation of biodiversity in Bangladesh. However, there is a risk in which valuable plant materials may be lost due to environmental hazards and high cost of their maintenance.

2.6.2 *Ex-situ* conservation

Ex-situ conservation, which includes conservation of samples representing genetic diversity away from own habitats. *Ex-situ* conservation is the chief mode of preservation for genetic resources, which may include both cultivated and wild materials, generally seeds or *in-vitro* maintained plant cells, tissues and organs which are preserved under appropriate conditions for long-term storage as gene banks. *Ex-situ* conservation is done by the establishment of gene banks, seed banks, botanical gardens, cultural collections etc. (Kumar and Mahendra, 2000). Baldah Garden, Wari, Dhaka and National Botanical Garden, Mirpur are the two best *ex-situ* conservation sites of Bangladesh. The collections of the Baldah garden is classified into seven categories. The garden has about 1200 plants species. Many of these are exotic and rare plants, perhaps the richest collection of *ex-situ* conservation of gene pools (Ahmed, 2003). National Botanical Garden established in 1961 as the store house of nearly 50,000 species of plants, herbs and shrubs including large collections of aquatic plants (Ahmed, 2003). Rare plant species are found in the garden such as white 'Rangan' (*Ixora superba*), little Mussaenda (*Mussaenda luteola*), white 'Chandan' (*Santalum album*) etc. (Ahmed, 2003). There are some Botanical Gardens in the campus of Dhaka University, Dhaka, Jahangirnagar University, Savar, and Bangladesh Agricultural University, Mymensingh. These gardens are also working as centres of *ex-situ* conservation of several rare and uncommon plant species collected from different ecological zones. Some research and academic institutions are also preserving gene pools and playing the role of *ex-situ* conservation.

CHAPTER 3: STUDY AREA

Bangladesh is situated in the north-eastern part of South Asia. It lies between $20^0 34'$ and $26^0 38'$ north latitude and between $88^0 01'$ and $92^0 41'$ east longitude. The country is bounded in the west, north and north-east by India. Towards the south-east, it has a border with Myanmar and faces the Bay of Bengal on the south. It encompasses an area of 147,570 square kilometers (sq. km). The present study was conducted in Patuakhali and Barguna districts under Southwestern zone of Bangladesh which were purposively selected to get desired information on homestead plant biodiversity of the region. These two districts are being affected by salinity intrusion and climatic disasters.

3.1 Patuakhali and Barguna districts

Patuakhali and Barguna districts of Southwestern region of Bangladesh fall under Barisal Division. These two districts lie between $21^0 40'$ and $32^0 36'$ north latitude and between $89^0 51'$ and $90^0 '40'$ east longitude. Patuakhai and Barguna are surrounded by Barisal district on the north, Jhalokathi and Bagerhat on the west and Bhola on the east. The Tetulia river, runs through the eastern side of Patuakhai and Barguna, falls into the Bay of Bengal separating these two districts from Bhola.

Patuakali and Barguna districts have a total area of 3,221 and 1,831 sq. km. with a total population of 15, 36000 (BBS, 2011). These two districts covered 3.42 % of the total area of Bangladesh. The average size of household of these two districts was 4.41 and 4.12, respectively. This coastal tidal flood plain area enjoys a number of diverse ecosystems.

3.2 Ecosystems of the study area

Bio-geographically, Bangladesh is situated in the "Oriental Region", lying in the transitional point between the Indo-Himalayans and Indo-Chinese sub-region of the Orient (Moran, 1984). The coastline of the country bounded by the Bay of Bengal is approximately 480 km in length. The Gangetic Tidal Flood plain constitutes about 49% of the coastal areas (SRDI, 2000). This area enjoys a number of diverse

16

ecosystems. However, in general, southern region of the country faces many impacts of climate change in the form of severe floods, cyclones, droughts, sea water level rise and salinity. The ecosystems of the coastal zone are delicate, dynamic and complex. Influencing factors of the ecosystems of the study zone are sea surge and waves, tides, water-logging, sedimentation, unplanned discharge and accretion etc. However, small-scale variation of physical, biological and socio-economic factors can also contribute enormously on plant diversity.

The major components of the ecosystem are the physical and chemical environment, the biotic elements and human interference. Each of these components influences the flora and the fauna, and in turn, these act on the environment. Ecosystems also affect qualitatively the nutrient cycle and life-supporting systems and inter and intra relations among the living entity (FRMP, 2000).

IUCN-Bangladesh has delineated Bangladesh into 12 bio-ecological zones; some of which are then further sub-divided into sub-zones. Accordingly, this delineation of greater Barisal and Khulna Division remained in the bio-ecological zones in the serial number 10 and 12, called Saline Tidal Floodplain and Coastal and Marine Water, respectively (Nishat *et al.* 2002). In total, nine districts of the coastal zone were included in the above-mentioned zones including Barguna and Patuakhali districts.

3.3 Characterization of the saline tidal floodplain and coastal marine water zones

3.3.1 Physiography

The land forming history and its genesis of the Saline Tidal Floodplain depicts a low ridge and basin relief, crossed by innumerable tidal rivers and creeks. Local differences in elevation are less than 1 m. The sediments are mainly composed of non-calcareous clay, although in the riverbanks, they are silt and slightly calcareous.

The coastal zone has its own dynamics and deserve special attention as very distinct terrain. The coastal area, comprising the complex delta of the Ganges-Brahmaputra-Meghna river system, has immense biological resources. The coastal morphology is influenced by high sediment accretion. Each year, about 2.4 billion tons of sediment is transported by the major rivers of Bangladesh having a profound effect on the geography of the floodplains and the coastal region, resulting in a net accretion of 35 km^2 of land per year (Rahman, 1992).

3.3.2 Rivers, canals and wetlands

Patuakhali and Barguna districts consist of a network of river system having a length of about 663.53 and 160.00 sq. km and their tributaries, which cover 20.71% and 22.00 %, of the areas of the districts. The different names of these rivers are given at different places. Rivers, tributaries, lakes and wetlands are the common source of surface water forming a "river system". Major rivers of the Patukhali district are Tetulia, Lohalia, Paira, Andharmanik, Agunmukha, Kajal, Burishar, Dhankhali, and the major rivers of Barguna district are Paira, Biskhali, Baleshar, Haringhata, Tiakhali, Behula etc. It has been estimated that about 39% of land area including char land, island and tributaries are in Patuakhali District. There are a good number of big chars which are Charmantaz, Charkashem, Charbiswas, Charkazal and Sonakatarchar, Fatrarchar, Laldiachar and Nisanbariachar belonging to Barguna district (Kamal and Rahman, 2005). Scarcity and abundance of water flow in these river systems is a common phenomenon that affects plant diversity.

3.3.3 Soils

The southwestern coastal zone consists of highland, medium highland and medium lowland. The soil of this southern region is clay and clay to sandy loam. The soil content of organic matter is poor due to over flooding and salinity. A substantial area becomes saline due to varying degrees of elevation in the dry season. Expansion of saline area during last three decades (between 1973 and 2000) is about 0.170 million hectares. The amount of salt accumulated is found more at the surface horizon.

3.3.4 Salinity level

Coastal saline area of the southern region has been gradually increasing in the recent past causing less cropping intensity with other severe physical and biological constraints. However, crop production and floral diversity in the salt affected areas of the coastal region differs considerably from the non-saline areas. Salinity causes unfavorable environment and hydrological situation restricting the normal crop production throughout the year. The degree of salinity varies widely with area and season, depending on the availability of fresh water, intensity of tidal flooding and nature of movement of saline ground water. It is indicated that about 1.02 million hectares, out of 1.459 million hectares of cultivated land in the coastal areas are affected by salinity of various degree (SRDI, 2000). Analysis of the soil salinity of some selected river water of Kalapara Upazilla under Patuakhali district has shown that water salinity level is about 31.0 dS/m in dry season (April-May). Accordingly, chemical composition of ground water (auger hole) at 2 sites at Kalapara of Patuakhali district is 24.0 and 29.0 dS/m, respectively.

3.4 Climate

Frequent natural calamities i.e. foods, droughts, cyclone etc are the result of climate change. It is a sensitive issue which affect agricultural and forestry production and livelihoods in Bangladesh. Climate change is the cause of unpredicted increased monsoon rainfall, temperature rise storm surge and other physical factors influencing overall production systems of the country.

3.4.1 Rainfall

Total rainfall of Patuakhali and Khepupara station under Patuakhali and Barguna districts, respectively, during 1983 to 2007 showed that rainfall level has been increasing but it fluctuates especially at Khepupara station (Figure-2). The average rainfall in the recent years was above 3000 mm. Bangladesh Climate Change Strategy and Action Plan (BCCSAP) 2008 stated that, "monsoon rainfall will increase, resulting in higher flows during the monsoon season in the river". Mandal, (2008) predicted that rainfall would become both higher and more erratic, and frequency and intensity of droughts are likely to increase.

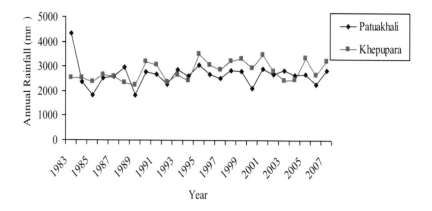

Figure 2. Total rainfall of Patuakhali and Khepupara stations under Patuakhali and Barguna districts during 1983 to 2007.

Source: Bangladesh Meteorological Department, Agargaon, Dhaka.

3.4. 2 Temperature

Temperature data of Patuakhali and Khepupara under Patuakhali and Barguna districts, respectively, during 1983 to 2007 showed that temperature has been increasing in recent years and it went up to 32^0 c in the year 2007 at Patuakhali station (Figure-4). According to prediction of Inter Governmental Panel on Climate Change (IPCC), global temperature will rise between 1.8^0 c and 4.0^0 c [by] the last decade of 21st century. These global warming will affect the mean sea level, especially the Bay of Bengal in the south of the country resulting an increased coastal flooding and saline intrusion into ponds, ditches, rivers and low-lying homesteads. The BCCSAP report stated that there will be increasingly frequent and severe floods, tropical cyclones, storm surges, and droughts, which will disrupt and displace millions of people from the coastal region, making them **"environmental refugees"**, unless existing polders are strengthened and new ones are built.

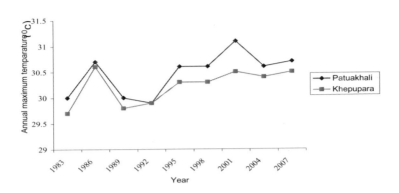

Figure-3. Temperature data of Patuakhali and Khepupara under Patuakhali and Barguna districts during 1983 to 2007.

Source: Bangladesh Meteorological Department, Agargaon, Dhaka.

3.4.3 Natural calamities

Bangladesh is ranked as the most vulnerable country to tropical cyclones and the sixth vulnerable country to floods (MoEF, 2008). In the last two decades, Bangladesh has experienced devastating floods and cyclones in 1970, 1991, 1998, 2000 and 2007. The cyclone of 1970, 1991 and super Sidr cyclone of 2007 were the worst, killed millions of people in the southern coastal zone and these are still memorable events that affected the livelihood and biodiversity of these areas.

3.4.4 Climate change and cyclone Sidr

The memorable Sidr cyclone of November 15, 2007, affected all southwestern districts which were: Patuakhali, Barguna, Pirozpur, Nalcity and Bagerhat. The centre of this cyclone is like an eye of man, so the meteorologists call it Sidr as presented in the Figure- 4.

Figure-4. Centre of the cyclone Sidr which affected the study area.

Sidr cyclone had originated from the Bay of Bengal area with 100 to 800 kilometers speed of wind ranging from 220 to 240 kilometers. About 80-90% homesteads and their houses were damaged. Sidr cyclone destroyed many lives and plants, preserved seeds and natural resources in the homesteads which influenced negatively in the livelihood of people. The damages of homestead plant resources will take a long time to fill the gap. A comprehensive step shall have to be undertaken to recoup the damage of plant diversity in homesteads, specific attention to be paid to naturally growing and threatened species in this region.

3.5 Bio-ecological characters of Southwestern coastal zone

A number of diverse ecosystems and their associated richness of plants have been documented. The utilization of homestead for coastal farming is an operational farm unit where crops, fisheries, livestock, poultry and multipurpose trees and shrubs are grown mainly for the purpose of satisfying farmers' basic needs.

3.5.1 Land use pattern

Because of poor land use intensity, many areas remained fallow and unused for the maximum time of the year. The major cropping patterns are:

i.   Fallow-Fallow-T. Amon: It is the common pattern practiced by the farmers.

ii. Fallow-T. Aus-T. Amon: It is the common, and major areas fall under this pattern.

iii. Rabi crops-Fallow-T. Amon: It is a moderate pattern in this area.

iv. Rabi-T. Aus-T. Amon: This pattern is practiced in lower areas.

3.6 Land tenure and leasing system

Land tenure and leasing system varies on land availability, local demand, capacity and cropping pattern of a particular zone. Different types of land tenure and leasing systems operated are described below:

i) **Owner cultivation:** Land is owned and operated by the owner. He himself took the risk of crop cultivation and selecting crops.

ii) **Temporary agreement (*Patta or Khaikalasi*):** This system is prominent and commonly used in which the land is mortgaged for 5-7 years on temporary contract basis for an amicably settled price. On expiry of the contractual period, the land virtually gets back to the land owner. This contract system is commonly practised for providing temporary needs, tackling any emergency and risk coverage etc.

iii) **Land agreement (money contact):** It is a long time mortgagee treated as semi-sale system. The land is mortgaged for particular amount of money for a definite period. If the land owner is able to pay back the money he will get back the land, otherwise the mortgager would enjoy the land as like as land owner. After the expiring of the contact period, if the original land owner fails to manage the mortgage money he is forced to sell the land. However, this is very rigid system which may turn original land owner into landless.

3.7 Land leasing system

Land is leased to other farmers on rental or on lease basis, usually for a limited period of time on payment of cash or kind. The price of land varies from place to place depending on production, cropping intensity and local demands etc. The following land leasing systems are being practiced:

i) **Temporary contract for one year:** Through this system landless farmers or agricultural labourers those who have no farmland make contact with the landowner for one year on the portion of land. They pay the land owner with a sum of fixed money depending on land type and productivity etc. Through this leasing system marginal class of farmers get access to land and have right to use the land and cultivate any suitable crops. But there are risks involved with occurring natural disaster which may damage the crops.

ii) **Share cropping:** Share cropping is a system of traditional agriculture in which a landowner allows a tenant to use a particular piece of land on a contract to share the produce of the land. The system of dividing crop into three shares: one for the landowner, one for the tenant and one for the input provider (landowner or tenant). Through this leasing system marginal class of farmers are losing their interest to cultivate land with the said one third returns.

iii) **Paddy contract:** In this process landowner and share cropper divide the paddy between two parties after harvesting crops. Part of paddy sharing depended on their

23

bilateral agreement based on input supports. In some cases tenant carry all harvested paddy at his own homestead. He (tenant) harvests the paddy for the landowner and sends it back to landowner house. A trust has been created between tenant and landowner. Through this process s

3.8 Hydrological constraint

There are two major hydrological problems, viz. in the winter season, most of the areas of this zone remain fallow during the months of January to June (a major constraint for the poor cropping intensity); the second is the monsoon tidal overflow which is uncontrollable by the existing faulty polder system. During the monsoon season, rain water congestion in the inside of the polder and overflow of water from the outside of the polder together cause inundation of the entire polder areas. This is why water is called "boon and bane" for the people of the coastal zone affecting homesteads plant biodiversity immensely.

3.9 Education

According to BSS, 2007, literacy rate of the study areas of Patuakhali and Barguna Districts was 51.65% and 55.28% higher in comparison to Bangladesh average of 46.15% (male of 50.26% and female of 41.79%). The literacy can influence the overall improvement of homestead's biotic resources. The educational status is better but economic condition of people of these districts is inferior in comparison to other regions of Bangladesh. People are adversely affected by recurring floods, cyclones, salinity and drought which are influencing low cropping intensity. Another consequence is rural to urban migration process where young people are migrating in large number.

CHAPTER 4: METHODOLOGY

In any scientific research, methodology plays an important role. Appropriate methodology enables the researcher to collect valid and reliable information to analyze the information properly to arrive at a correct conclusion. The present study was accomplished by collecting primary data using survey method. The information was collected and supplemented by secondary information to enrich the study findings. The information was collected and compiled following the standard procedures and methods of observation, repeated visits to the study areas and use of structured questionnaires.

4. 1 Area of the study

The study area was Southwestern zone of Bangladesh and it was seriously affected by salinity intrusion and climate change. Patuakhali and Barguna of two deltaic districts of Southwestern zone were purposely selected as the study area based on the salinity level. These districts consist of 7 and 5 administrative upazillas including 882 and 563 villages, and 280980 and 180060 households, respectively (BSS, 2002).

4. 2 Salinity level of the study sites

The salinity status of the study area was obtained from secondary data of Soil Resources Development Institute (SRDI). According to SRDI report coastal saline area of Bangladesh are classified into three areas: less saline (salinity ranged from 2.1 to 8.10 dS/m), moderately saline (salinity ranged from 8.1 to 12.0 dS/m), strongly saline (salinity ranged from 12.1 to 16.0 dS/m. or > 16.0 dS/m). Side by side, salinity intensity categorized into four classes were S_1, S_2, S_3 and S_4. Based on these data, salinity intensity of 11 upazillas of Patuakhali and Barguna districts of the study were assessed (Figure-5). It was found that the strongly-saline area was Kalapara (75%) and the less-saline area was Bauphal (17%) upazilla of Patuakhali district. On the other hand, Amtoli (64%) upazilla of Barguna district was moderately-saline area (SRDI, 2000).

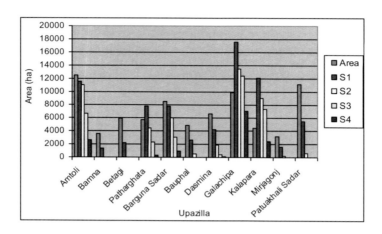

Figure-5. Salinity status of various upazillas belonging to Patuakhali and Barguna Districts.

4. 3 Site selection

The study was carried out in three upazillas of these two peripheral districts (Patuakhali and Barguna) of the Bay of Bengal. The upazillas were Bauphal, Amtoli, and Kalapara. These three upazillas were selected based on the level of salinity i.e., strongly saline, moderately saline and less saline or slightly-saline site. Kalapara was highly-saline site, Amtoli moderately saline site, and Bauphal less saline site (Map-2). Out of these three upazillas, one union and one village from each upazilla were selected as sampling areas. The selected unions were Latachapali (Kalapara), Karaibaria (Amtoli), and Kalisuri (Bauphal); and three villages were Nayapara, Choulapara and Kalisuri, respectively (Table-1). The salinity classes of SRDI (S_1, S_2, S_3 and S_4) was used to select the study villages. The selected villages were encompassed at least one canal or a small river. The village transact is shown in Appendix-1.

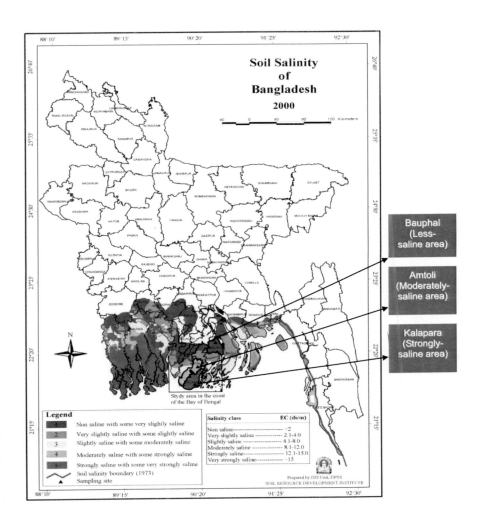

Map-2. Map showing the selected three upazillas having different types of salinity level of the study sites under the two peripheral districts facing Bay of Bengal.

Table-1. Sampling sites (Villages, Unions, Upazillas, and Districts) as per the salinity level of the study areas.

Village	Union	Upazilla	District	Salinity parameter	Salinity level
Nayapara	Latachapali	Kalapara	Patuakhali	Strongly Saline	12.1-16 and >16 ds/m
Choulapara	Karaibaria	Amtoli	Barguna	Moderately Saline	8.1-12 ds/m
Kalisuri	Kalisuri	Bauphal	Patuakhali	Less Saline	<2 ds/m

4.4 Procedure of data collection

The major processes of data collection were: i) Formal survey (enumeration and household data collection) and ii) Discussion meeting. The other supporting process of data collection included the collection of secondary information from different organizations and interactions with local experienced persons.

4.5 Questionnaire development

According to the objectives of the study, a structured questionnaire was prepared (both open and close type of questions) to collect the required data on various aspects of homestead plant biodiversity, utility of the products and income generation. A set of pre-tested structured questionnaire (Appendix-2) was prepared for data collection through face-to-face interviews. The questionnaire was translated into Bengali for easy access to systematic conversation and information collection.

4.6 Survey

The information of this study was accomplished through survey. Two types of information were collected through survey. These were: i) enumeration of the homestead plants and ii) collection of household information. Head of the household either male or female had provided the information. In some homesteads, household female head and children also joined in the interview process which had improved the

28

quality of data and minimized error. The interviewer was carefully conducted the interview after the devastating Sidr of 15 November 2007, which affected the study areas. However, the people became surprised to observe the enumeration process of homestead plants. They passed comment in their own language, "*Badara o-hane kare-ki, gach gone! Etagula gach gonte parbe? O Allah ! manush gonte dekchi, gach gonte dehinai gibane*". "What they were doing over the homesteads? O ! God, we have seen population census, but did never see counting of plants in life."

4.7 Time of data collection

The information was collected from the four different farm categories of households during July 2006 to March 2008. The devastating Sidr of 15 November 2007 had hampered the study and also lengthened the time frame for the study.

4. 8 Sampling method

Based on the farm size, the respondent households were categorized into four different groups, i.e., i) large household (>2 ha), ii) medium household (1.01-2.0 ha), iii) small household (0.51-1.0 ha) and iv) marginal household (0.21-0.51 ha).

4.8.1 Sample size and sampling

A multistage sample tool was used to identify household and the tribes of the Village. Verifying and combining the available baseline household survey of DAE and Union Council, a list of households was prepared as the sample population of the study. The families living outside the embankment were excluded from the quick survey. The total number of households in the villages were also obtained from the District Census published by the BBS, 2001, where total household of the villages were recorded Nayapara 129, Choulapara 196 and Kalisuri 756. Considering the larger size of Kalisuri Village, it was divided into two sub-blocks as per DAE working procedure. During the study, a sub-block was selected adjacent to the Aloki river. Therefore, total households of Nayapara, Choulapara and Kalisuri was 157, 209 and 305, respectively. In total, 36% of proportionate sample was drawn from the total population size 671. The proportionate distribution of sample households were Landless 131 (35%), Small 330 (29%), Medium 160 (34%) and Large 50 (60%). Total sample of study was 240 of which 80 samples were drawn from different household for each location.

4.9 Types of data collection

Different types of information were collected through questionnaire survey. The major types of data were as follows:

i.   Household demographic information

ii.   Structure of homestead and utilization of land

iii.   Homestead plant species identification and characterization

iv.   Economic and profitable and multipurpose tree species

v.   Seed collection and preservation techniques

vi.   Seedlings plantation and felling trend of trees

vii.   Management practices of homestead production

viii.   Problems faced by the household

ix.   Gender role in homestead plant management and conservation

x.   Household income and expenditure

xi.   Food security and relative contribution of trees, vegetables, and on-farm, off-farm activities

xii.   Loss/degradation and conservation of homestead plant species

4.10 Relative prevalence and species diversity indices

The following two formulae were used to measure relative prevalence, species diversity indices and equitability.

4.10.1 Relative prevalence of species

To indicate the importance and richness of different plant species in study areas, relative prevalence (RP) of species was calculated as follows:

RP = Population of the species / homestead X % homesteads with the species.

Relative prevalence of all types of trees was calculated by using the above formula.

4.10.2 Shannon-Wiener species diversity index

To measure the abundance and diversity of different plant species among the different farm category and saline zones, species diversity indices were calculated by using Shannon-Wiener (H) and Simpson's (D) formula.

Species diversity index: To ensure the abundance and diversity of different plant species, Shannon-Wiener Species Diversity Index (H) as shown below was used:

$$H = - \sum (Pi \ LnPi)$$

Where, Pi is the proportional abundance of ith species such that $Pi = n/N$ (n is the number of the individuals in ith species and N is the total number of the individuals of all species in the community).

4. 10.3 Simpson's species diversity index

Simpson's Species Diversity Index (D).

$$D = 1 - \sum_{i=1}^{S} (Pi)^2$$

Where P_i is the proportion of total individuals in the ith species.

$P_i = n/N$, n is the number of individuals in the ith species and N is the total number of the individuals of all species in the community, D=Diversity index number, $\sum$= is a summation sign, S=Total number of species, D can range from 0 to1.

4.10.4 Equitability

Equitability means equality or evenness. Diversity index depends on richness and equitability. Equitability can itself be quantified by expressing Simpson's diversity index (D) as a proportion of maximum possible value D would assume if individuals were completely evenly distributed among the species.

In fact, D max = S

The Equitability,

$$E = \frac{D}{D\ max}$$

$$= \frac{1}{\sum\limits_{i=1}^{s} pi^2} \times \frac{1}{s}$$

$$= \frac{D}{S}$$

Equitability assumes a value between 0 and 1.

4.10. 5 Correlation among the seven variables

By variant correlation was done by using SPSS-10 soft ware.

4.10. 6 Types of species planted

Percentage of individual farmer's category was calculated on the basis of total seedlings planted/ homestead.

4.10. 7 Felling trend of trees

Felling trend was determined for the different farm categories according to different plant species and age distribution based on total felled trees.

4.10. 8 Household income and expenditure

Household income and expenditure was distributed in three categories and expenditure in seven categories. Proportion of individual item (income and expenditure) was calculated on the basis of total income and expenditure as follows.

Percent of income generated from individual item = taka earned from that individual item /total income of a farmers x 100.

Percent of an individual expenditure = total expenditure for that item/total taka expenditure x 100.

4.11 Determination of economic and profitable fruit-timber and vegetable-yielding species

The most economic and profitable timber and fruit-yielding and diversified vegetable-yielding species were selected on the basis of income and uses of these species. The species were weighed by applying value 1 for the lowest priority and 10 for the highest priority through which economic and profitable species were listed. The economic and profitable timber-yielding species evaluated as many uses are: i) timber, ii) fuel wood, iii) fodder, iv) furniture, v) industrial uses, vi) handicrafts, vii) pole and other uses, viii) agricultural tools and other uses, ix) market trade/sell value, x) medicinal/herbal uses. Similarly economic and profitable fruit-yielding species evaluated as many uses are: i) fruit/food, ii) juice, iii) market price/cash-money, iv) medicinal/herbal uses value, v) industrial uses, vi) fodder/bird food, vii) small-business/trade, viii) jam/jelly/pickles, ix) risk coverage, x) house/boat making/handicrafts etc.

Vegetable species grown in the homesteads were recognized with diversified uses categorized into: i) use of leaves, ii) fruit value, iii) use of stem/whole body of plant (stem and roots as vegetables), iv) medicinal value, v) tuber/stem/bark for smash, vi) use of tuber/flower as vegetables, vii) cash/income generation, viii) preparing jam/jelly/pickle and salad, ix) direct consumption of unripe fruit, x) social and other cultural value etc.

4.12 Determination of multipurpose tree species

Profitable multipurpose timber and fruit-yielding species identified which is almost common and have provided economic return and supply food, fuel, fodder and nutrition supply to the household. The trees are not used as mono-purpose, although grows for more than one or multipurpose uses. Their priority ranking was fixed and arranged according to the choice of the respondents. Total of 15 uses of major species were identified. These trees served as multipurpose species. The multipurpose uses were categorized into: i) fruit, ii) timber, iii) fuel wood, iv) medicinal uses, v) fodder, vi) furniture, vii) molasses/juice, viii. industrial uses, ix) handicrafts, x) pole and other uses, xi) agricultural tools and other uses, xii) jam/jelly/pickles, xiii) animal/bird food, xiv) market value, xv) hedge/roof and thatching materials. According to multifarious uses some of the species were prioritized through ranking and weighed (putting value 1 for the lowest priority and 15 for the highest priority).

4. 13 Discussion meeting

Focus Group Discussion (FGD) was held in each survey area to know the existing homestead economic plant species, utilization, and contribution to their livelihoods. Local eminent persons including public representatives, school teachers and old experienced farmers were contracted to discuss the ranges of issues regarding homestead biodiversity and future management options. Discussion meetings were held with relevant government and non-government organizations working in the study areas for verifying the collected information.

4.14 Secondary information collection

Secondary information were collected from the different organizations such as Bangladesh Meteorological Department, Department of Forest and Environment (DOEF), Department of Agriculture Extension (DAE), Soil Resource Development Institute (SRDI), National Herbarium, Bangladesh Agricultural Research Council (BARC), SAARC-Agriculture Information Centre (SAIC), Jahangirnagar University and Bangabandhu Sheikh Mujibur Rahman Agriculture University (BSMRAU) libraries, and Non-Government Organizations.

4.15 Data analysis

Microsoft Excel Program was used to process all collected information and in preparing chart and graphs. SPSS (Statistical Package for Social Science) software was used to estimate the descriptive statistics of the data.

CHAPTER 5: RESULTS AND DISCUSSION

The findings are presented in the following sections as per objectives of the study in the sequence of demographic and land profile, homestead plant identification, utilization and conservation.

5.1 Household demographic and land profile

Demographic and land profile of the households determine the quality of life, influence on homestead plant biodiversity and play role in ensuring household food security. Selected demographic information of the respondents are presented in this section.

5.1.1 Gender, marital and religion of the respondents

The distribution of the respondents based on gender, marital and religion are presented in Table-2. Among the respondents, 5.42% were female and 94.58% male which indicated that male members are dominating the family. Marital status of the respondents was grouped into three categories as married, unmarried and widow. The highest proportion of the respondents was married (95%), followed by unmarried (2.5%) and widow (2.5%). It was found that some women in the society were treated as widow although they were not divorced, because their husbands left them and got married with another woman. These deserted women are becoming vulnerable in the society. In view of religion, 87.50% of the respondents were Muslim and 12.50% belonged to Hindu and Rakhain (Mogh) communities.

5.1.2 Age of the respondents

The respondents were grouped into three categories based on their ages (Figure-6). The results were that young, middle age and old age categories of the respondents were 24.6%, 50.8% and 24.6%, respectively. The middle age group represented more than half of the respondents which included a representative part of the farmers of the society.

Table-2. Distribution of the respondent based on gender, marital and religion at the studied sites of the southwestern coastal region of Bangladesh.

Farm category	Gender		Marital status				Religion	
	Male	Female	Married	Unmarried	Widow	Total	Muslim	Other**
Landless	44	4	46	1	1	48	42	6
Small	97	5	97	2	3	102	92	10
Medium	58	2	55	3	2	60	49	11
Large	28	2	30	00	00	30	27	3
Total ( )*	227 (94.)	13 (5.42)	228 (95.0)	6 (2.5)	6 (2.5)	240 (100)	210 (87.50)	30 (12.50)

* Figures in the parentheses indicate the percentages of the respondents.

** Hindu and Rakhain (Mogh) community.

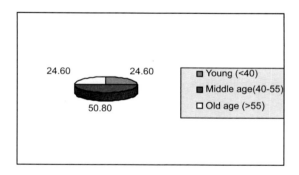

Figure-6. Age category of the respondents (%).

5.1.3 Education level of the respondents

The respondents were grouped into five education levels as shown in Figure-7. More than half of the respondents (52%) belonged to primary education followed by secondary education (34%), higher secondary education (8%), above secondary education (3%), and only 3% was illiterate. Secondary information of the study sites

36

revealed that the adult literacy rates of the southern Barguna (53.6%) and Patuakhali (51.6%) districts were higher than that of the national average (46.3%) (BBS, 2006). The higher literacy rate in the study area might be favorable for improving the household decision making and livelihood standard.

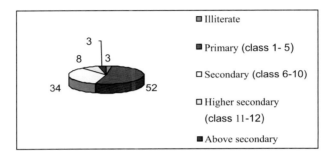

Figure-7. Education levels of the respondents (%) of the studied southwestern coastal region of Bangladesh.

5.1.4 Household size of the respondents

The household size of the respondents is presented in Table-3. The average household size was 5.35 in the study area, which was higher than the national average of 4.90 (BSS, 2002). The average size of dwelling household of Barisal division was 5.40 (BSS, 2002) which was almost similar to the study areas. It was found that 4.07 persons were adult and 1.27 under-adult.

Table-3. Household size and sex distribution of the respondents among the different categories of the household.

Farm category	Family size	Male	Female	Adult	Under-adult
Landless	5.15	2.79	2.33	3.81	1.31
Small	5.27	2.81	2.49	3.88	1.35
Medium	5.43	3.00	2.43	4.30	1.15
Large	5.80	2.97	2.83	4.63	1.17
Total	5.35	2.88	2.49	4.07	1.27

5.1.5 On-farm and Off-farm occupation of the respondents

The occupation of the respondents was broadly categorized into on-farm and off-farm which is shown in Table-4. Respondents were involved both on-farm and off-farm occupations. Majority of the respondents (97%) were involved in on-farm occupation, while only 3% were landless farmers. Besides, the highest proportion (63.33%) of the household families was involved in off-farm occupation of which landless were (22.37%), followed by small (43.42%), medium (23.68%) and large (10.53%) families. Off-farm occupation, especially for the marginal farmers was found to rise in the rural areas which included variety of small business, service, driving, boating, handicraft, marketing, post-harvesting, fish processing etc.

Table-4. On-farm and off-farm occupations of the respondents of the studied southwestern coastal region of Bangladesh.

Farm category	Respondents occupation	
	On-farm	Off-farm
Landless	42	34
Small	101	66
Medium	60	36
Large	30	16
Total	233 (97.08)	152 (63.33)

* Figures in the parentheses indicate the percentages of the respondents.

5.1.6 Training received by the respondents

Training is an important tool for enhancing knowledge and skill to show change in behaviors of the respondents in respect of homestead production and biodiversity conservation (Table- 5). It was observed that 12.9 % respondent farmers received formal or informal training and 87.1% didn't receive any training. Government Organizations (GOs), such as Department of Agriculture Extension (DAE), Bangladesh Rural Development Board (BRDB) and Department of Forest (DoF) were mainly involved in organizing formal or informal training particularly targeted for the middle-grouped and large-grouped framers. On the other hand, Non-Government Organizations (NGOs), such as Bangladesh Rural Advancement Committee (BRAC), Grameen Bank (GB) working in the studied areas were involved in the operation of credits compared to knowledge and technology development. Atikullah and Hassanullah (2007) reported that NGOs have been spending more time and effort for credit management rather than technical know-how improvement of the farmers. However, majority of the farmers of all classes, especially landless and small were lagging behind to acquire knowledge in agro-sylviculture and biodiversity improvement. Ahmed (1999) assessed farmers response to training needs and found that 80% of the farmers would like to undertake training with special attention to homestead management, species and site selection, disease identification, pest control, tree improvement etc.

Table-5.Training received by the respondents of the studied southwestern coastal region of Bangladesh.

| Farm category | Respondents | | Total |
	Received training	Not received training	
Landless	2 (0.8)	46 (19.2)	48 (20.0)
Small	15 (6.29)	87 (36.25)	102 (42.5)
Medium	9 (3.75)	51 (21.25)	60 (25.0)
Large	5 (2.08)	25 (10.42)	30 (12.5)
Total	31 (12.92)	209 (87.08)	240 (100)

* Figures in the parentheses indicate the percentages of the respondents.

5.1.7 Land holdings of the respondents

Land holding of the respondents is presented in Table-6. Total land holdings of the respondents increased as the farm size increased. Landless farmers hold only 0.23 ha of land per farm and large farmers hold 14.67 ha of land per farm which proved maximum land resources were controlled and managed by the large farmers. Average homestead size of landless farmers was 0.18 ha which was about 8 times higher to large farmers (1.59 ha per farm). However, average homestead size (0.42 ha) of the coastal region of the study was bigger than that of other regions of Bangladesh. These bigger homesteads offer an unique opportunity for producing more outputs and making the households more economically solvent. Homestead is the main source of landless farmers; therefore, they used to pay more attention to on-farm and off-farm activities for their survival and many other socioeconomic activities.

Table-6. Household land holdings of the respondents of the southwestern coastal region of Bangladesh.

Farm category	Respondent farm land (ha)		
	Homestead land /farm	Cultivated land/farm	Land /farm
Landless	0.18 (78)	0.05 (22)	0.23 (100)
Small	0.22 (37)	0.37 (63)	0.59 (100)
Medium	0.46 (16)	2.38 (84)	2.84 (100)
Large	1.59 (11)	13.08 (89)	14.67 (100)
Average	0.42 (15)	2.42 (85)	2.84 (100)

* Figures in the parentheses indicate the percentages.

5.1.7.1 Land ownership patterns of the respondents

Land is a finite and valuable resource upon which farmers depend for food, fiber and fuel and the basic amenities of rural life. Five major categories of ownership patterns were identified in the study areas (Table-7). The majority (70.83%) of lands were owned through inheritance from the parents followed by purchase (38.75 %), and rentals (14.17%). A 5% of the respondents were the recipients of Government fallow ('khas') land by which some families had been settled in this region.

Table-7. Land ownership patterns of the respondents of the surveyed southwestern coastal region of Bangladesh.

Ownership pattern	Respondent		Ranking
	Number	Percentage	
Parental	170	70.83	1
Purchased	93	38.75	2
Rentals	34	14.17	3
Government fallow (Khas)	12	5.00	4
Others	02	0.83	5

5.2 Configuration and space utilization of homestead

Configuration and space utilization pattern of the homesteads varied according to ecological zones and farm categories. There were many interrelated attributes that affect the decision making, land arrangement, utilization of resources and homestead plant biodiversity. In the following sections these attributes, such as homestead configuration, status of homestead space utilization and pattern of homestead land utilization are discussed.

5.2.1 Homestead configuration

Homesteads facing at different configuration are presented in Table-8. A good portion of the respondents (41.70%) constructed their homesteads at South direction followed by East (34.60%) and West directions (14.20%). Among the farm categories i.e., large, medium, small and marginal farmers preferred to construct homesteads in the South direction were 50%, 48.3%, 34.3% and 43.8%, respectively. Generally farmers learnt from the Khanna's proverbs, "South facing homestead is to be considered as superior homestead". Therefore, well-planned homesteads by using biotic and abiotic attributes encompass tree and vegetable to be increased productivity of homesteads.

Reasons were identified and listed while constructing the homesteads with different directions which determine land use, production, communication, environment and also social factors. The reasons were prioritized as per their choice as: i) homestead linked-up with village road/main road and with embankment, ii) open air flow from the Bay of Bengal and for open space (available sunlight, less cold and less disease infection), iii) ease for homestead plantation and cultivation as well as safe movement of women, iv) homesteads linked-up with rivers, canals and ease for transportation of agricultural products, and v) homesteads established by their ancestors or old-aged people which followed their liking and desire.

Table-8. Homestead construction with different directions of the southwestern coastal region of Bangladesh.

Homestead configuration	Distribution of homesteads (number and percent)				Total
	Farm category				
	Marginal	Small	Medium	Large	
East direction	16 (33.3)	39 (38.2)	15 (25.0)	13 (43.3)	83 (34.6)
West direction	8 (16.7)	15 (14.7)	11 (18.3)	0.0	34 (14.2)
North direction	03 (6.3)	13 (12.7)	05 (8.3)	02 (6.7)	23 (9.6)
South direction	21 (43.8)	35 (34.3)	29 (48.3)	15 (50.0)	100 (41.7)

* Figures in the parentheses indicate the percentages of the respondents.

5.2.2 Homestead space utilization

Homestead space is utilized for diversified purposes (Table-9). Out of 240 sample homesteads, 63 (26.25%) were over-utilized, 40 (16.66%) fully utilized, 113 (47.08%) partially utilized and the rest were 24 (10.0%) under-utilized that left fallow. More than half of the homesteads (57.08% of homesteads consisting 47.08% medium and 10.0% under-utilized) have potential of using maximum land for increasing

41

production. Basher (1999) observed the spatial arrangement of the homesteads space utilization as 15% over-crowded, 55% fully utilized and 21% under-utilized. Rahim and Haider (2000) find out that multi-layered tree garden or multi-storied cropping is practiced in and around the homesteads of Bangladesh in an unsystematic manner. There was a substantial part of land is found for improvement of homesteads through properly managed space planning which can alleviate poverty. Rahim and Haider (2000) also reiterated that multi-storeyed tree/cropping system in the homesteads can increase income substantially from per unit area. Proper use of technical knowledge and improve cropping practices for homestead space utilization to be introduced is a need of the study.

Table-9. Level of homestead space utilization at the studied sites of the southwestern coastal region of Bangladesh.

Farm category	Level of homestead space utilization (number and percent)				Total
	Over Utilized	Full utilized	Partially Utilized	Under utilized	
Landless	17 (27.0)	8 (20.0)	20 (17.7)	3 (12.5)	48 (20.0)
Small	23 (36.5)	17 (42.5)	54 (47.80)	8 (33.3)	102 (42.5)
Medium	16 (25.4	5 (12.5)	30 (26.5)	9 (37.5)	60 (25.0)
Large	07 (11.1)	10 (25.0)	09 (8.0)	4 (16.7)	30 (12.5)
Total	63 (100))	40 (100)	113 (100)	24 (100)	240 (100)

* Figures in the parentheses denote the percentages of the respondents.

5.2.3 Pattern of homestead space utilization

Homestead space arrays in a pattern for the combined production of vegetables, fruits, timber, fish, poultry and livestock products (Figure-8). The results of the study represent that the highest area of the homesteads (29.42%) were used for gardening purpose. In addition to these gardening area, pond dykes, approach road, and surrounding area of the homestead were also used as garden along with agro-sylviculture pattern. The second highest portion of land was occupied by ponds and pond dykes (27.67%), followed by houses including poultry shade (14.0%), surroundings area and raised borders (12.49), court yard (11.21%) and approach road (5.21%) etc. More over collectively, the gardens and surrounding area occupied with 41.91% area. Basak (2002) reported a pattern of homestead space utilization in which trees and bushes occupied 53% and houses 19.0%. A common and unique tendency was found that farmer's prepared usually big sized court yard, houses and excavated mini pond for multipurpose production along with many other social utilization. Besides, a common shade prepared as a live fence with trees and shrubs across the raised surrounding line (*kandi*) of the homestead may be envisioned as a wind breaker and fence (*parda*) for women and other social purposes. The homestead's spatial and multi-storeyed space utilization has turned it into a center of socio-economic activities and developed as a production system where habitats can live as well survive with an eco-friendly atmosphere.

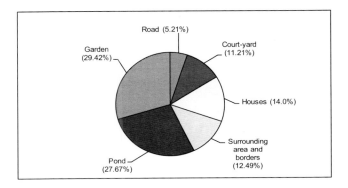

Figure-8. Pattern of homestead space utilization in the study areas of the south-western coastal region of Bangladesh.

5.3 Species diversity and characterization

The species found in the homesteads were identified and documented in this section. The plant species has been listed as per categories of uses such as fruit-yielding, timber-and fuel-yielding, medicinal, ornamental, naturally growing, and non-woody (herbs/shrubs/climbers) etc which are briefly mentioned below:

5.3.1 Plant species diversity

A total of 189 plant species were identified from 240 homesteads of 3 study sites i, e., less saline, moderately saline and strongly saline areas (Table-10). The number and kind of plant species grown in the three sites varied and it was 189 in moderately saline area, followed by 167 in strongly-saline area and 173 in less saline area. Out of 189 plant species, 67.20% were trees (timber, fruit, medicinal, ornamental and naturally growing) and 32.80% were non-woody (herb/shrub/climber) types. The highest number distributed in the moderately-saline areas were fruit-yielding species 40, timber and fuel wood species 36, medicine and spices species 17, ornamental species 20, naturally growing plant species 14 and other species. These identified

43

plant species are listed (Appendix-3) along with their scientific names, family names, and local and English names as well as their economic uses.

Alam and Masud (2005), Millat-e-Mustafa and Haruni (2002) identified a total of 142 species and 162 species from homestead, respectively. The study revealed that species diversity is relatively higher than other parts of the region. A few number of plant species were totally disappeared in less saline area such as Sundri (*Heritiera fomes*). This variation in different saline zones may be multipurpose species along with some other saline loving fruit-yielding and timber-yielding species were found to grow well in saline ecosystem.

The study find out species diversity and richness is relatively higher than other parts of the region. The large number of different plant species found in the southern coastal region of Bangladesh showed richness of plants in terms of organismal diversity. This could be due to these region enjoys rich as well unique ecosystem and tidal sediments which is influencing factor of rich plant diversity.

Table-10. Homestead plant species diversity in different salinity zones of the studied southwestern coastal areas of Bangladesh.

Homestead plant species	Different saline zones of the study areas		
	Less saline	Moderately saline	Strongly saline
Fruits-yielding[1]	39	40	40
Timber and fuel-yielding[2]	32	36	34
Medicinal and spices[3]	15	17	14
Ornamental plant species [4]	16	20	13
Naturally-growingplant species [5]	13	14	14
Nonwoody[6]. (herbs/shrubs/climbers)	58	62	52
Total	173	189	167

-Fruit-yielding species[1], Timber and fuel-yielding species[2], Medicinal and Spice-yielding plant species[3], Ornamental plant species[4], Naturally growing plant species [5], Non-woody plant species [6].

5.3.2 Species composition

Species composition in homestead on the basis of salinity level is shown in Table-11. On an average, about 181 species existed per homestead which was encouraging in terms of homestead plant biodiversity. In Table-11, it could be seen that the highest number (201.26) of trees existed per homestead in less saline zone, followed by 176.30 in highly saline zone and 164.08 in moderately saline zone. In the less-saline zone, multipurpose species were grown because the species were adapted there. Even some fruit-yielding and timber-yielding species were found to grow well in less saline zones of the study areas.

Table-11. Tree composition per homestead on the basis of salinity level in the studied areas of the southwestern coastal region of Bangladesh.

Tree species	Number of tree species/homestead of different saline zones			Species/ Homestead
	Less saline	Moderately saline	Strongly saline	
Timber-yielding	122.28	92.14	97.90	104.10
Fruit-yielding	74.26	66.41	71.66	70.78
Medicinals	1.15	3.29	3.41	2.62
Ornamentals	0.86	1.16	1.05	1.03
Naturally-growing	2.71	1.08	2.28	2.02
Total	201.26	164.08	176.30	180.55

5.3.3 Tree density

Tree density on the basis of farm size of different farm categories is shown in Table-12. On an average, a total of 343 number of trees existed in a homestead having a size of one acre land. Among the farm categories, landless farmers accommodated the highest number of plants (553). On the other hand, large farmers accommodated less number of plants (207) in comparison to their homestead size since they have kept some space in their homesteads for other agricultural purposes.

Table-12. Tree density on the basis of farm size of different farm categories of the studied southwestern coastal region of Bangladesh.

Farm category	Respondents' land holdings		Tree density (number/acre)
	Total farm land (acre)	Homestead land (acre*)	
Landless	26.98	21.67	553
Small	149.14	39.37	433
Medium	420.55	67.7	317
Large	1087.17	117.7	207
Total	309.81	52.7	343

- Hectare = 2.47 acre.

5.4 Identification of homestead's economic and profitable tree species

Plants and vegetable species in the context of economic value, profitability and multipurpose uses are summarized. Some vegetables and fruit species were identified for food security along with support during crisis period especially for the pro-poor.

Gggggggggggggggggggggggggggggggggggggggggggggggggggggggggggggggggggggggggg

5.4.1 Major economic and profitable timber-yielding species

Collectively homestead trees species was found to play diversified roles in supporting of food, income and safety net. The economic and profitable timber-yielding species characterized and listed in the study areas are presented in Figure-9. In total 11 species were earmarked as economic and profitable timber-yielding species. The species were: i) *Samanea saman*, ii) *Albizia richardiana*, iii) *Swietenia macrophylla*, iv) *Albizia procera*, v) *Bambusa tulda*, vi) *Azadirachta indica*, vii) *Pongamia pinnata*, viii) *Neolamarckia cadamba*, ix) *Pithecellobium dulce*, x) *Terminalia catappa*, and xi) *Cassia fistula*. Among the identified species, Raintree followed by Chambol and Mahogony were common and prioritized as the top ranking economic and profitable timber-yielding tress across the study areas. Considering the economic return and safety net these species were treated as the poor man's species for **"poverty reduction"**. These species provided rapid economic return to the households particularly rural poorer. Other species which were common in saline areas were Koroj, Bamboo, Country neem, Indian buch, Jilapi, Kadam, Indian almond and Indian laburnum. These species also adapted and performed well for their quick growth and economic return. It was used as an economic and profitable species for diversified purposes as fuel, pole and timber etc. Jilapi is a timber yielding species but provides fruits for the rural children, adolescents and youths which are an alternative source of nutrition in the south-central zone of Bangladesh.

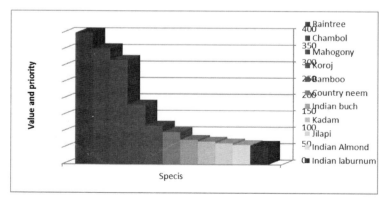

Figure-9. Economic and profitable timber-yielding species in the south-central coastal zone of Bangladesh.

46

5.4.2 Economic and profitable fruit-yielding species

A total of 14 economic and profitable fruit-yielding species were identified and prioritized as per their economic value and profitability in coastal area (Figure-10). These species were: i) *Mangifera indica*, ii) *Cocos nucifera*, iii) *Zizyphus mauritiana*, iv) *Borassus flabellifer*, v) *Artocarpus heterophyllus*, vi) *Psidium guajava*, vii) *Tamarindus indica*, viii) *Areca catechu*, ix) *Musa* sylvestris, x) *Citrus aurantifolia*, xi) *Sygygium fruticosum*, xii) *Spondias pinnata*, xiii) *Diospyros blancoi*, and xiv) *Citrus maxima*. It was observed that Mango, Coconut, Banana, Betel Nut, Bilati gab, Hog palm grow and perform well in Bauphal area; and Mango, Jujube, Coconut, Palmyra palm, Pummelo, Tamarind, and Lemon in Kalapara and Amtoli areas. The fruit-yielding species play an important role as a source of major fruits for the households. The present investigation found that Mango, Coconut, Palmyra palm, Banana, Tamarind and Jujube should be considered as economic and profitable species. According to farmers opinion a marginal family can ensure part of their annual income and livelihood from these fruit-yielding species. The leaves and fruits of some of these species (Coconut, Palmyra palm)were reported to be utilized for multipurpose uses, such as molasses, juice, handicrafts, shade, mat, and small-sale business as well seasonal employment creation for the rural landless and marginal people.

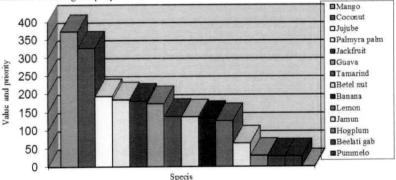

Figure-10. Economic and profitable fruit-yielding species in the south-central coastal zone of Bangladesh.

5.4.3 Multipurpose tree species

Tree species having multipurpose uses and characterized based on their various utilization aspects are presented in Table-13. A good number of woody perennial tree species were grown around the homesteads served as multipurpose species. Out of them 20 species were identified and listed as multipurpose species according to their manifold uses that varied from 14 to 8. Species scored the highest ranking in terms of their uses were: Mango (14 uses) followed by Jackfruit (13 uses), Palmyra palm (12 uses), Tamarind (11 uses), and Betel nut (10 uses). This woody species have wider ranges of uses in the homesteads to fulfill farmer's multifarious daily needs. These species also were found to perform as multi-storied production in a complex agroforestry system (Rahim and Haider, 2000). A key advantage of multipurpose trees, however is that it provides benefits to for poor farmers especially for Bangladesh (ICRAF, 2010). Therefore, it was concluded that multipurpose timber-fruit yielding species should be given preference for plantation and sustainable production and overall development of species diversity in the homesteads.

Table-13. Multipurpose tree species uses, weightage and ranking in the study areas of the south-central coastal zone

Scientific name	English name	Local name	Uses	Weightage	Ranking
Mangifera indica	Mango	Aam	14	119	1
Artocurpus heterophyllus	Jackfruit	Kathal	13	105	2
Borassus flabellifer	Palmyra palm	Tal	12	94	3
Tamarindus indica	Tamarind	Tetul	11	93	4
Diospyros blancoi	Wood nut	Beelati gab	11	89	6
Areca catechu	Betel nut	Supari	10	88	7
Alstonia scholaris	Devils tree	Chatian	10	83	7
Albizia procera	Siris	Karoi	11	87	8
Syzygium fruticosum	Jamun	Deshi jam	10	86	9
Cocos nucifera	Coconut	Narikel	10	86	9
Albizia richardiana	Chapalish	Chambol	10	77	11
Bambusa tulda	Bamboo	Bash	10	77	11
Acacia nilotica	Arabic gum	Babla	11	76	12
Erythrina fusca	Coral tree	Kata mandar	11	76	13
Pithecellobium dulce	Jilapi	Jilapi	10	75	14
Dillenia indica	Indian dillenia	Chalta	09	74	15
Ficus hispida	Country fig	Dumur	08	72	16
Cassia fistula	Indian laburnum	Sonail	08	69	17
Ceiba pentandra	Kapok-tree	Kat tula	08	62	18
Lannea coromandelica	Wadier	Kapila/ Jiga	08	60	19

5.5 Vegetable species diversity in different seasons and salinity level

A little variation was found in the number and kind of vegetable species grown in the different saline zones of the study areas. A total of 53 vegetable species was found to grow in less saline area, 56 in moderately saline area, and 55 highly saline areas (Table-14). *Moringa oleifera* (Sajna) and *Acrostichum aureum* (Tiger fern) are totally absent in less saline area specially these vegetables is highly suitable in highly saline to moderately saline zones. The seasonal distribution of vegetables were possible to classify as: 20 (35.71%) year round, 20 (33.71%) summer-grown, and 16 (28.58%) winter-grown. The number of year round vegetables species was found higher in different saline zones which is good indication for growing vegetable cultivation in these areas. *Vigna unguiculata* (Felon), *Zea mays* (Bhutta), and *Cucumis sativus* (Khiroi) are not known as vegetables but these were seen to be cultivated in a few homesteads as an alternative source of subsistence food. Few vegetables are specially liked by the "Rakhain", an aboriginal community traditionally built their habitat in highly to moderately saline areas in this region. These vegetables were *Hibiscus sabdariffa* (Chikur), *Moringa oleifera* (Sajna), *Sesbania grandiflora* (Bakphul), *Solanum torvum* (Titbegun), and *Acrostichum aureum* (Tiger fern) etc. These vegetables were found as of bitter and sour taste due to which the Rakhain community used them in their food recipe.

Table-14. Seasonal distribution of vegetable species in different saline zones of the southcentral coastal region of Bangladesh.

Vegetable growing season	Vegetable grow in different saline zones of the study			Total (%)
	Less saline	Moderately saline	Strongly saline	
Year round	17	20	19	20 (35.71%)
Summer	20	20	20	20 (33.71%)
Winter	16	16	16	16 (28.58%)
Total	53	56	55	Total (100 %)

5.6 Species characterization

Specific characters of the species were verified by which the existing plants were systematically identified for evaluating the variation of species.

5.6.1 Systematics of the plant species

The identified plants were systematically classified according to their species, genera and families are given in Table-15. The available growing plant species were belonged to 74 families, 152 genera and 189 species which included a wider range of diversity of dicot and monocot plants. Similarly, Arefin *et al.* (2011) identified 72 families and 183 genera under from the Satchari National Park. It was found the family, Euphorbiaceae represented by the highest 9 genera and 10 species, followed by Poaceae (8 genera and 9 spp) and Fabaceae (6 genera and 8 spp) are mark as the major families in the study sites.

Tabl-15. Systematics (families, genera and species) of the homesteads' plant species of the studied areas of the southwestern coastal region of Bangladesh.

Family	Genus	Species	Family	Genus	Specie s	Family	Genus	Species
1 Acanthaceae	3	3	26 Crassulaceae	1	1	51 Myrtaceae	3	6
2 Amaranthaceae	3	3	27 Cruciferae	1	1	52 Nyctaginaceae	1	1
3 Amaryllidaceae	1	1	28 Cucurbitaceae	1	1	53 Oleaceae	1	1
4 Anacardiaceae	3	4	29 Cyperaceae	2	2	54 Pandanaceae	1	1
5 Angiopteridaceae	1	1	30 Dilleniaceae	1	1	55 Poaceae	8	9
6 Annonaceae	2	3	31 Dioscoreaceae	1	2	56 Polygonaceae	1	1
7 Apiaceae	1	1	32 Ebenaceae	1	2	57 Pontederiaceae	1	1
8 Apocynaceae	4	4	33 Elaeocarpaceae	1	1	58 Portulacaceae	1	1
9 Araceae	1	1	34 Euphorbiaceae	9	10	59 Punicaceae	1	1
10 Arecaceae	6	7	35 Fabaceae	6	8	60 Rhamnaceae	1	1
11 Asclepiadaceae	2	2	36 Flacourtiaceae	1	1	61 Rosaceae	2	2
12 Asteraceae	2	2	37 Flagellariaceae	1	1	62 Rubiaceae	5	5
13 Averrhoaceae	1	1	38 Lamiaceae	1	2	63 Rutaceae	3	7
14 Bignoniaceae	2	2	39 Lauraceae	2	3	64 Sapindaceae	2	2
15 Bombacaceae	2	2	40 Lecythidaceae	1	1	65Sapotaceae	2	2
16 Boraginaceae	2	2	41 Liliaceae	2	2	66Scrophulariaceae	1	1
17 Brassicaceae	1	1	42 Lythraceae	1	1	67 Solanaceae	2	2
18 Bromeliaceae	1	1	43 Magnoliaceae	1	1	68 Sonneratiaceae	1	2
19 Caesalpiniaceae	4	4	44 Malvaceae	2	3	69 Sterculiaceae	2	2
20 Cannaceae	1	1	45 Marantaceae	1	1	70 Tiliaceae	1	1
21 Casuarinaceae	1	1	46 Meliaceae	5	6	71 Typhaceae	1	1
22 Chenopodiaceae	1	1	47 Menispermaceae	1	1	72 Verbenaceae	5	6
23 Clusiaceae	2	2	48 Mimosaceae	6	10	73 Vitaceae	1	1
24 Combretaceae	1	4	49 Moraceae	3	8	74 Zingiberaceae	2	4
25 Convolvulaceae	2	3	50 Moringaceae	1	1	Subtotal =24	49	61
Sub total =25	50	57	50	53	71	Grand total =74	152	189

5.6.2 Systematics of the vegetable species

These vegetable species were systematically classified according to their family and genera (Table-16). The vegetable species of the study area were found to belong to 41 genera of 21 families. It indicates the presence of a wide diversity of vegetable yielding plants in this region. The family Cucurbitaceae and Febaceae represented by eight species ranked top of the list followed by, Solanaceae and Araceae. Eight families belonged and represented by only one species. It indicated the existence of a relatively less available vegetable yielding species was needed to care properly for their conservation and regeneration.

Table-16. Systematic arrangement of homestead vegetable grown in the south-central coastal region of Bangladesh.

Family	No of Genera	No of Species	Family	No of Genus	No of Species
Acrostichaceae	1	1	Dioscoreaceae	1	1
Amaranthaceae	2	3	Fabaceae	6	8
Araceae	3	5	Malvaceae	1	2
Asteraceae	2	2	Moringaceae	1	1
Athyriaceae	1	1	Musaceae	1	2
Basellaceae	1	1	Nymphaeaceae	1	1
Caricaceae	1	1	Polypodiaceae	1	1
Chenopodiaceae	2	2	Solanaceae	3	6
Convolvulace	1	3	Tiliaceae	1	1
Cucurbitaceae	6	8	Umbelliferae	3	3
Cruciferae	2	3			
Total= 11	22	30	21	41	56

5.6.3 Characterization of tree species based on salinity level

Characterization of tree species based on salinity level or saline-tolerant species is difficult. However, a total of 29 species were identified under moderate to strongly saline areas and 19 species were for less-saline areas (Table-17). The salinity intrusion and drought, especially in the dry season were common in these areas that caused serious damage to homestead production of both fruit-and timber-yielding species; hampered income generation and food/nutrition of the households. This information would be useful for increasing saline tolerant genetic race and increasing over all production of this region.

Table-17. Tree species grown in moderate to strongly saline tolerant and less saline areas of the study sites of the southwestern coastal region of Bangladesh.

Local name	Scientific name	Local name	Scientific name
Aam[1]	*Mangifera indica*	Kadam[2]	*Neolamarckia cadamba*
Akashmoni[1]	*Acacia auriculiformis*	Kailla lata[1]	*Derris trifoliata*
Amra[2]	*Spondias pinnata*	Kala[2]	*Musa sylvestris*
Atafal[2]	*Annona reticulata*	Kamranga[1]	*Averrhoa carambola*
Babla[1]	*Acacia nilotica*	Kathal[2]	*Artocarpus heterophyllus*
Bahai/Dumur[1]	*Ficus hispida*	Kaufal[1]	*Garcinia cowa*
Boroi[1]	*Zizyphus mauritiana*	Kewra[1]	*Sonneratia apetala*
Bash[2]	*Bambusa tulda*	Keya pata[1]	*Pandanus foetidus*
Bel[1]	*Aegle marmelos*	Khejur[2]	*Phoenix sylvestris*
Beelati gab[1]	*Diospyros blancoi*	Jilapi[1]	*Pithecellobium dulce*
Chaila[1]	*Sonneratia caseolaris*	Lebu[2]	*Citrus lemon*
Chambol[1]	*Albizia richardiana*	Lichu[2]	*Litchi chinensis*
Chatian[2]	*Alstonia scholaris*	Mahogony[1]	*Swietenia macrophylla*
Choto Jam[1]	*Syzygium fruticosum*	Mama kala[1]	*Sarcolobus carinatus*
Dalim[2]	*Punica granatum*	Narikel[1]	*Cocos nucifera*
Deshi neem[1]	*Azadirachta indica*	Jhau[1]	*Casuarina littorea*
Golpata[1]	*Nypa fruticans*	Pepey[2]	*Carica papaya*
Harguji[1]	*Acanthus illicifolius*	Pechi gab[1]	*Diospyros malabarica*
Jambura[2]	*Citrus maxima*	Peyara[2]	*Psidium guajava*
Jmrul[2]	*Feronia elephantum*	Raintree[1]	*Samanea saman*
Safeda[2]	*Maniekara zapota*	Sarbati lebu[2]	*Citrus aurantium*
Sundari[1]	*Heritiera fomes*	Sarifa[2]	*Annona squamosa*
Supari[2]	*Areca catechu*	Sissoo[1]	*Dalbergia sissoo*
Tal[1]	*Borassus flabellifer*	Tetul[1]	*Tamarindus indica*

* [1] Saline tolerant to moderately saline tolerant and [2] less saline tolerant species.

5.6.4 Characterization of vegetable species based on salinity level

Some vegetable species were reported to grow in moderately saline to less-saline areas in the homesteads of the coastal zone (Table-18). The identified vegetable species were: Eryngium, Winged bean, Sword bean, Indian spinach, Turnip, Tomato, Drumstick, Tiger fern, Roselle, Bakphul, Country bean, Sweet potato, and Radish. Sweet potato as a popular vegetable of this region also grows well in the homesteads of less-saline areas of the coastal zone. Saha (2001) also reported that some species grow well in moderately saline or less saline areas. These were Coriander, Tomato, Country bean, Radish, Potato, Bush bean etc. According to the respondents, they need more saline-tolerant and heavy rain tolerant species for increasing homestead vegetable cultivation.

Table-18. Vegetable-yielding species grown in moderate to strongly saline tolerant and less saline areas of the studied sites of the southwestern coastal region of Bangladesh.

Local name	Scientific name	Local name	Scientific name
Bilatidhania[1]	*Eryngium foetidum*	Sajna[1]	*Moringa oleifera*
Kamranga seem[1]	*Psophocarpus tetragonolobus*	Oudhachopa[1]	*Acrostichum aureum*
Mouseem[1]	*Canavalia gladiata*	Chikur[1]	*Hibiscus sabdariffa*
Buno kakrol[1]	*Momordica cochinchinensis*	Bakphul[1]	*Sesbania grandiflora*
Puisak[2]	*Basella rubra*	Seem [1]	*Lablab purpureus*
Shalgom[1]	*Brassica campestris*	Mistialu[1]	*Ipomoea batatas*
Tomato[2]	*Lycopersicon lycopersicum*	Mula[1]	*Raphanus sativus*

* [1] Moderately saline-tolerant, and [2] Less- saline-tolerant.

5.6.5 Low cost minor fruits and vegetables

Some low cost minor fruits and vegetables were found to grow well in this region (Table-19). These vegetables are rich in nutritional value and popular among the poor. The species were: Seeded banana, Plantain banana, Eryngium, Bush bean, Indian almond, Banana, and Tamarind. These species have wider range of economic importance to the farmers in the context of this region, while these species are decreasing because of socio-economic and environmental factors. These species have adaptive ability in the southwestern region of Bangladesh which need to be promoted for cultivating in wider scale for proper utilization and conservation.

Table-19. Low cost minor fruits and vegetable-yielding species grown in the study areas of the southwestern coastal region of Bangladesh.

Local name	English name	Scientific name
Aita kala	Seeded banana	*Musa sylvestris*
Anaji kala	Plantain banana	*Musa sapientum* var. *paradisiaca*
Bilati dhania	Eryngium	*Eryngium foetidum*
Felon*	Bush bean	*Vigna unguiculata*
Katbadam	Indian almond	*Terminalia catappa*
Kathali kala	Banana	*Musa paradisiaca*
Tetul	Tamarind	*Tamarindus indica*
Choila	---	*Sonneratia caseolaris*
Cau phal	Cowea	*Garcinia cowa*
Jilapi	Jilapi	*Pithecellobium dulce*
Pechi gab	River ebony	*Diospyros malabarica*
Bombai marich	Hot chilli	*Capsicum frutescense*

* Felon is a pulse crop, but in some homesteads it is also grown to be used as vegetable (smash) in immature form.

5.6.6 Rare and threatened species

Some rare and threatened species of fruit, vegetable and shrubs were recorded through this investigation. The species were identified to be a few in number at the homesteads as single or few individuals. Some of the species were found as threatened or would be endangered over time. The local, English and scientific names along with their family status of these species are shown in Table-20.

Table-20. Rare and threatened species of the homesteads in the study areas of the southwestern coastal region of Bangladesh.

Local name	Scientific name	Family	Status
Unknown	*Hyptis capitata*	Labiatae	Threatened
Abeti	*Flagellaria indica*	Flagellariaceae	Threatened
Bantula	*Hibiscus moschatus*	Malvaceae	Threatened
Buno Kankrol	*Momordica cochinchinensis*	Fabaceae	Threatened
Dakur	*Cerbera odollam*	Apocynaceae	Less available
Kamranga sheem	*Psophocarpus tetragonolobus*	Fabaceae	Lees available
Mamakala	*Sarcolobus carinatus*	Asclepidaceae	Rare
Mewa kathal	*Annona muricata*	Annonaceae	Less available
Mamakathal	*Morinda angustifolia*	Rubiaceae	Less available
Mouseem	*Canavalia gladiate*	Fabaceae	Lees available
Nagmani	*Wissadula periplocifolia*	Malvaceae	Less available
Pechigab	*Diospyros malabarica*	Ebenaceae	Less available
Sada dhutra	*Datura metel*	Solanaceae	Rare
Urigab/Bangab	*Diospyros montana*	Ebenaceae	Rare

5.6.7 Dominant species in the homesteads of Hindu and Rakhain communities

Hindu community in the study area has been preserving some species of ornamental and vegetable-yielding plants at their homesteads, such as, Basil (Tulsi), but it was not found in Muslim homesteads. Similarly some vegetables are being cultivated by the "Rakhain" community in their homesteads, particularly in strongly and moderately saline areas. These included Roselle (Chukur), Drumstick (Sajna), Bakphul, Titbegun, and Ouddachopa (Tiger fern) etc. They grow those species because they are fond of bitter and sour taste for which they use them in their food recipes. List of some dominant species in Hindu and Rakhain communities is given in Table-21.

Table-21. Dominant species in the homesteads of Hindu and Rakhain communities of the southwestern coastal region of Bangladesh.

Local name	English name	Scientific name	Uses
Bakphul	----	*Sesbania grandiflora*	Vegetable, fence
Chukur	Roselle	*Hibiscus sabdariffa*	Medicine, Puja
Ganda	Marigold	*Tagetes patula*	Medicine, Puja
Kalke/Karali	Lucky nut	*Thevetia peruviana*	Medicine, Puja
Oudhachopa	Tiger fern	*Acrostichum aureum*	Curry, Frying
Sajna	Drumstick	*Moringa oleifera*	Medicine, Fruit
Tit begun	Bitter brinjal	*Solanum torvum*	Vegetable
Tulsi	Basil	*Ocimum basilicum*	Medicine, Puja

5.7 Relative prevalence of plant species

Relative prevalence is importance to measure species richness of different plant species in varying saline zones of the study areas.

5.7.1 Relative prevalence of tree species

Relative prevalence is importance to measure species richness of different plant species of the study area. Relative prevalence of the most dominant species is presented in Figure-11. Out of listed tree species 15 prevalent species was listed in this figure. Most prevalent and top ranking timber-yielding species were *Albizia richardiana*, *Swietenia mahagoni* and *Samanea saman*; fruit-yielding species were *Mangifera indica*, *Cocos nucifera* and *Phoenix sylvestris*; medicinal and spice-yielding species were *Azadirachta indica*, *Terminalia arjuna* and *Calotropis gigantea*; ornamental species were *Lawsonia inermis*, *Delonix regia* and *Hibiscus rosa-sinensis* and naturally growing species were *Streblus asper*, *Hydnocarpus kurzii* and *Ficus hispida*. Significant variation in number of tree species and variation among the categories (timber-, fruit-, medicinal and spice-yielding, ornamental and naturally growing) of plant species in the homesteads can be observed. Relative prevalence of other species (apart from the top prevalent species) was very poor. This is an indication of diminishing trend of homestead plant organismal and ecological plant diversity in terms of number of different local plants species in the homesteads.

The top prevalent species in the homesteads were Chambol (28.01%), Mahogony (26.67%) and Raintree (22.68%), Mango (10.88%), Coconut (10.06%), and Date palm (6.96%). Only a few top prevalent timber-yielding species is increasing whereas those of other categories such as medicinal, ornamental and naturally-growing tree species in the homesteads were being gradually replaced. This is an alarming and an indication of valuable plant species erosion from the homesteads. The south-central homesteads were observed rich in fruit-yielding species, but the relative prevalence of that local and indigenous species had been decreasing notably. Therefore, it is essential to promote policy guidelines and motivate farmers to maintain a combination to grow local fruit-yielding species along with other species to encompass different types of trees in the homesteads. Otherwise, homestead plants diversity will be eroded over time. Relative prevalence of species was listed and as found in Appendix-4.

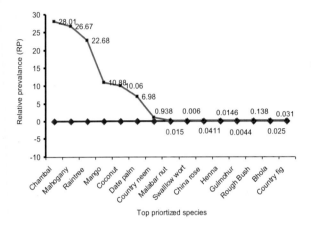

Figure-11.Relative prevalence of dominant species (timber-yielding, fruit-yielding, medicinal, ornamental and naturally growing) in the studied sites.

5.7.2 Relative prevalence of major species at various saline zones

Relative prevalence of major timber-yielding and fruit-yielding plants in varying saline zones has been presented in Figure-12. It was observed that Chambol and Mahogony were highly dominant in less saline areas but Raintree was dominant in moderately and strongly saline areas. In case of fruit-yielding species, Mango was prevalent in highly saline areas; while Coconut was almost equally dominant in all the saline zones of the study areas. However, Date palm was less dominant in highly saline areas compared to moderately and less saline areas. It implied that species adaptation in different sites widely varied due to different levels of salinity.

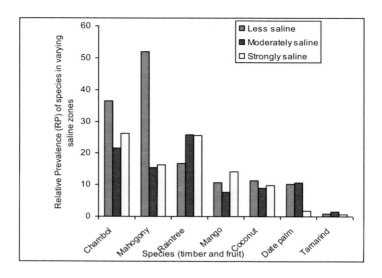

Figure-12. Relative prevalence of major timber-and fruit-yielding species in different saline areas of the studied southwestern coastal zones of Bangladesh.

5.7.3 Tree species reduced to a few individuals in the homesteads of the study areas

Some of the species that existed in one or more homesteads were (Appendix-5) reduced to a fewer number of plants in all farm categories which could not show their relative prevalence values. This posed a question regarding their existence in terms of genetic and organismal diversity. The species, which were reduced to a few individuals, were considered as threatened species. These species are getting reduced from the homesteads due to various reasons, such as heavy felling of Kadamba (*Neolamarckia cadamba*) and Jamun (*Syzygium fruticosum*) for economic value, less attention to transplant seedlings and non-availability of seedlings such as Kadamba (*Neolamarckia cadamba*), Haritaki/Black myrobalan (*Terminalia chebula*), Sajna/Drumstick (*Moringa oleifera*), Ulotkambal/Delvil's cotton (*Abroma augusta*) and less awareness about the species. On the other hand, some species are also being tried to be introduced such as Sarbati lebu (*Citrus aurantium*), Jalpai/Indian olive (*Elaeocarpus robustus*), Kadbel/Elephant apple (*Feronia limonia*) which are unable to adapt or survive due to salinity. Punnal/Gaitta (*Calophyllum inophyllum*) is a salt-tolerant species, but it is decreasing from the homesteads due to its multipurpose uses like agricultural tools, boat making and fuel wood. However, to fill the gap of these less prevalent species necessary steps are to be taken for *in-situ* and *ex-situ* conservation. Species showing no relative prevalence values owing to their fewer number of individuals are listed along with their local/English and scientific names of some species as shown in Table-22.

Table-22. Species existed in one or more homesteads but reduced to a few individuals in the study areas of the southwestern coastal region of Bangladesh.

Local name	Scientific name	Local name	Scientific name
Bahal	*Cordia dichotoma*	Kadbel	*Feronia limonia*
Bain	*Avicennia officinalis*	Kewra	*Sonneratia apetala*
Chaitain	*Alstonia scholaris*	Mewa kathal	*Morinda angustifolia*
Jag dumur	*Ficus glomerata*	Nauasonail	*Oroxylum indicum*
Gaitta	*Calophyllum inophyllum*	Royna	*Aphanamixis polystachya*
Haritaki	*Terminalia chebula*	Sajna	*Moringa oleifera*
Jalpai	*Elaeocarpus robustus*	Sarbati lebu	*Citrus aurantium*
Jam	*Syzygium fruticosum*	Ulat kambal	*Abroma augusta*

5.7.4 Tree species of very low prevalence value

Relative prevalence value of some species (Appendix-5) ranged from 0.001 to 0.03 is presented in Table-22.1. These species may be reduced in number from the homesteads over time. This is an alarming sign of decreasing homestead plants, especially which are uncommon, indigenous and minor in the southwestern coastal zone of Bangladesh. The species prevalence value at a very low level may be considered as rare species.

These species are already reduced from the homesteads due to various reasons, such as heavy felling, salinity, and less awareness about their importance. Since Sundri, was common in the homesteads of this region, but now a days farmers are preserving a few of them because of their liking towards the species of economic value and for household use. Wild cinchona, has been felled in large number from this region due to its use in match factory. This species naturally grows from seed. Raising of its seedlings is technically difficult. Hence it is a great problem to fill the gap of this species. Acacia, as an exotic species, has been introduced in the homesteads and for planting by the road sides and embankment. On the other hand, some other species are also being introduced such as Litchi and Sapota, but these could not survive due to salinity. In order to fill the gap of these species in the southern homesteads a comprehensive plan is needed for massive plantation, initiated by raising salt-tolerant cultivars of these commercially important fruit-yielding species.

Table-22.1 Tree species belonging to very low relative prevalence values at the southwestern coastal region of Bangladesh.

Local name	English name	Scientific name	RP (all farm)
Amloki	Indian gooseberry	*Phyllanthus emblica*	0.01
Atafol	Bullock heart	*Annona reticulata*	0.01
Buno amra	Wild hogpalm	*Spondias pinnata*	0.002
Chaila	----	*Sonneratia caseolaris*	0.003
Chalta	Indian dillenia	*Dillenia indica*	0.01
Cou	Cowa	*Garcinia cowa*	0.01
Dalim	Pomegranate	*Punica granatum*	0.01
Gigni	---	*Trema orientalis*	0.002
Jhau	Seef wood	*Casuarina equisetifolia*	0.0003
Lebu	Lemon	*Citrus aurantifolia*	0.01
Kadam	Wild cinchona	*Neolamarckia cadamba*	0.01
Litchu	Litchi	*Litchi chinensis*	0.01
Pahari neem	Neem	*Melia azedarach*	0.001
Pakur/Pipar	Pipal	*Ficus religiosa*	0.01
Mandar	Coral tree	*Erythrina veriegata*	0.01
Sonail	Indian laburnum	*Cassia fistula*	0.02
Sundari	Sundri	*Heritiera fomes*	0.01

5.7.5 Relative preference of vegetable species

Relative preference values of vegetable species is presented in Appendix-4, however, the top most preferable vegetable species are presented in Table-23. The top most preferable vegetable grown well in the homesteads was *Lagenaria siceraria*, followed by *Carica papaya*, *Lablab purpureus*, *Cucurbita maxima*, *Luffa acutangula*, *Amarnathus tricolor*, *Trichosanthes anguina*, *Cucumis sativus*, *Solanum melongena*, and *Basella rubra*. These vegetables were reported to grow in the homesteads using various indigenous cultivation methods and support system such as raised bed, pit (mada), roof top, house corner and cow-shed. Beside these vegetables, other vegetables do not grow well due to heavy rain in the monsoon, drought and also due to salinity in the dry season.

Table-23. Top ranking most preferred vegetable species grown in the studied southwestern coastal region of Bangladesh.

English name	Local name	Scientific name	Number of homesteads cultivating vegetables	RP (%)
Bottle gourd	Lau	*Lagenaria siceraria*	146	60.83
Papaya	Pepey	*Carica papaya*	139	57.92
Country bean	Sheem	*Llablab purpureus*	125	52.08
Pumpkin	Mistikumra	*Cucurbita maxima*	120	50.00
Ribbed gourd	Jhinga	*Luffa acutangula*	92	38.33
Red amaranth	Lalsak	*Amaranthus tricolor*	89	37.08
Snake gourd	Rekha	*Trichosanthes anguina*	86	35.83
Cucumber	Sashsa	*Cucumis sativus*	83	34.58
Egg plant	Begoon	*Solanum melongena*	81	33.75
Indian spinach	Puisak	*Basella rubra*	81	33.75

5.8 Diversity indices of different categories of species

The simplest measure of the character of a community that takes into account both the abundance pattern (evenness/equitabily=E) and the species richness (S) by Simpson's diversity index (D) and the Shannon-Wiener diversity index (H) are presented for different farm categories of the respondents in this section. D and H diversity indices of the species were calculated as per formula discussed in Chapter-4 and in Sections 4.10.2 and 4.10.3.

The diversity indices of different species under different categories is presented in Table-24. Table showed that the diversity (D) and abundance were the highest in large farm (0.8613) and decrease gradually as the farm size decreased in which diversity values in medium, small and landless farm categories were 0.8483, 0.8355 and 0.8379, respectively.

The similar trend of variation was found in case of diversity (H) where the value was the highest for large farm (2.36). This gradually decreased to medium (2.27), small (2.22) and landless (2.21) farm categories. The comparison between two methods of biodiversity indices (D and H) showed positive relation in the measurement of diversity indices. On the other hand, there was a positive correlation between plant diversity and farm categories, viz. it can be concluded that plant diversity increased as the farm size increased.

Fruit-yielding species: Side by side, population of fruit-yielding species followed a negative trend among the farm size in which the fruit trees diversity increased as the farm size decreased. It also showed the same sequence as reported by Basak (2002). These consequently proved that landless and marginal farmers were usually biased on fruit-yielding species for household consumption, for food security and for sale as a way of income generation and to meet the household daily expenditure.

Timber-yielding species: Table-23 showed that the diversity and abundance of timber-yielding species of all farm categories was higher than that of fruit-yielding, medicinal, ornamental and naturally growing species. On the other hand, the variation of diversity and abundance of fruit-yielding species is less than timber-yielding species in farm categories. The distinguished difference between timber-yielding species and fruit-yielding species of small farm was 0.11 and 0.05. This variation may have negative relationship in terms of species and ecosystem diversity which in the long run may create negative impact in the livelihood and life cycle of all the inhabitants. Diversity index and equitability (E) of trees according to farm category was worked out as per Simpson's formula. Equitability values varied in different farm categories in which large farm (0.36) was followed by medium (0.33), small (0.30) and landless (0.30). There is a positive correlation between diversity indices (D) and equitability (E) where diversity indices decreased with equitability along with the diminishing size of the farm categories.

Table-24. Species diversity indices and equitability of different farm categories of the studied areas of the southwestern coastal zone of Bangladesh.

Farm category	Pi^2					D*	E	H**
	Timber-yielding	Fruit-yielding	Medicinal	Ornamental	Naturally growing			
Landless	0.1112	0.0495	0.001	0.0001	0.0004	0.8379	0.3022	2.21
Small	0.1117	0.0517	0.0005	0.0001	0.0003	0.8355	0.3088	2.22
Medium	0.1066	0.0433	0.0014	0.0001	0.0003	0.8483	0.3323	2.27
Large	0.0983	0.0396	0.0004	0.0001	0.0003	0.8613	0.3625	2.36

*D=Simpson's diversity index, E =Simpson's equitability and ** H = Shannon-Wiener index, P_i = population of total individuals in the ith species.

5.8.1 Diversity indices of different categories of species in different saline zones

The diversity indices differed in various saline zones of the study areas (Table-25). The highest diversity index was found in moderately saline area (0.8727), followed by strongly saline (0.8395) and less saline areas (0.815). The same trend was noted in the diversity indices of Shannon-Wiener (H) formula. Diversity indices of fruit-yielding species (0.04) in less saline zone was a little bit higher in comparison to other zones, while the naturally growing species (0.0005) showed a little bit higher value in saline areas than other zones. The plausible reasons of variation in the biodiversity indices among the areas in which the homesteads of moderately saline areas adapted saline tolerant, moderately saline tolerant as well as less saline species which could have positive influence to increase the biodiversity indices.

Table-25. Species diversity indices of different categories of species in different saline zones of the study areas of the southwestern coastal region of Bangladesh.

Salinity class	Pi^2					*D	E	**H
	Timber-yielding	Fruit-yielding	Medicinal species	Ornamental species	Naturally growing species			
Less saline	0.1366	0.048	0.0001	0.0001	0.0004	0.815	0.2551	2.11
Moderately saline	0.0831	0.0424	0.0015	0.0001	0.0002	0.8727	0.3825	2.36
Strongly saline	0.1064	0.0526	0.001	0.0001	0.0005	0.8395	0.3226	2.28

*D=Simpson's diversity index, E =Simpson's equitability and ** H = Shannon-Wiener index, P_i = population of total individuals in the i th species.

5.8.2 Correlation between age, education, family size and availability of different plant species

Age, education, timber-yielding and fruit-yielding species the respondents had positive relationship with their total species (Table-26). All categories of farmers were equally interested to plant timber-yielding species, followed by fruit-yielding species, but educated farmers preferred timber-yielding species which had positive correlation with young-aged and small families. A unique correlation was found where educated farmers had shown likeness to plant and grow medicinal and ornamental plants at their homesteads. This is a good indication of conservation of plant diversity in homesteads. As a whole, abundance of timber trees followed by fruit trees in the homesteads of south-central region of Bangladesh was correlated with the high value of species diversity index.

It reveals that age and education not correlated in which old aged respondents were less educated and young aged respondents were relatively educated. At the same time, the family size of an aged respondent was bigger in comparison to a young farmer.

This was due to awareness increasing among the young farmers and they are taking more responsibility to maintain their family. It is essential to motivate the young and educated farmers to enhance the plant diversity in homesteads for increasing income and regular food supply for the family members. This information was also helpful to increase food and nutrition security of the households of the studied area.

Table-26. Correlation between age, education, family size and different plant species of the studied areas of the southwestern coastal region of Bangladesh.

Variables	Correlation value					Total species
	Timber-yielding species	Fruit-yielding species	Medicinal species	Ornamental species	Naturally growing species	
Age	0.106	0.147*	-0.066	-0.001	0.103	0.135*
Education	0.102	0.147*	0.153*	0.185**	-0.064	0.137*
Family size	0.004	0.153*	-0.078	-0.09	-0.034	0.051
Timber-yielding species		0.43	-0.042	-0.009	0.028	0.949**
Fruit-yielding species			0.1	0.094	0.087	0.69**
Medicinal species				0.398**	0.036	0.037
Ornamental species					0.046	0.05
Naturally growing species						0.077

**Correlation is significant at the 0.01 level.
*Correlation is significant at the 0.05 level.

5.9 Changing pattern of growing plant species

Trend of changing pattern of growing different plant species in the homesteads was calculated to understand the changes in species over time in the southwestern coastal region.

5.9.1 Changing pattern of growing plant species over time

Tree species status in the homesteads over time was compared between present (2008) and 20 years ago (1988), and shown in Figure-13. Twenty years ago the number of fruit-yielding species was higher compared to others species. At that period, fruit-yielding species status was 32.0 which was higher than timber-yielding species (23.4). In the Figure-13 it could be seen that fruit-yielding species increased gradually from

1988 to 2008. On the other hand, number of timber-yielding species during the same time increased remarkably. The number of timber-yielding trees jumped to 104 in 2008. This figure proved that homesteads of the southwestern coastal region dominated with timber-yielding species over time. Among the timber-yielding plants, only three to five species were relatively more prevalent. This may be the consequence of ever-increasing demand of timber and fuel wood for the region. The relative prevalence and diversity index of timber-yielding species were also higher in this region that was proved in the previous section. Traditionally, this region is known to be predominant in indigenous fruit-species. However, the changing status and variation of species over time may create impediment on the overall improvement of plant biodiversity in the region. Side by side, this gradual increasing trend of fruit-yielding species over time may create negative impact in the production, nutrition and income generation as well as livelihood of all the farmers of the study areas in the long run. A noticeable trend was found in the medicinal plants where the number decreased over time from 13 to 3. Viewing the naturally growing and ornamental plants status in the homestead, the same diminishing trend was found from 26 to 2.0 and 8 to 1.02 over time. This indicated that this valuable naturally growing species are being felled for multi-purpose use. Considering the overall changing pattern of growing timber and fruit-yielding species, the existing number is not conducive in terms of plant biodiversity. It is essential to look forward immediately to protect fruit-yielding, medicinal and naturally growing species for their utilization and conservation.

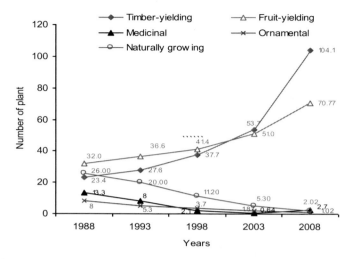

Figure-13. Changing pattern of growing of different plant species in the homesteads of the studied areas of the southwestern coastal region of Bangladesh.

5.9.2 Tree species per homestead over time

Existing trees per homestead over time are presented in Figure-14. The findings of the study was that irrespective of type of species, number of species having the age of 10 years per homestead was higher than that of the number of species having the age of above 11 years and up to 20 years. This gap was more predominant in case of timber-yielding species as compared to other species. This indicated that households are planting higher number of timber-yielding plants during the last 10 years because of an expectation for a rapid economic gain. However, increase in the number of timber trees per homestead having the age of below 10 years was 96, while it was above 8 having the age of 11 and up to 20 years. On the other hand, in case of fruit-yielding trees, the difference was less where number of fruit trees having the age below 10 years was about 46 and it was 25 having the age above 11 to 20 years. The proportion of timber and fruit trees in each homestead was not balanced which must have harmful effect for overall homestead production and maintenance for sustainable ecosystem.

66

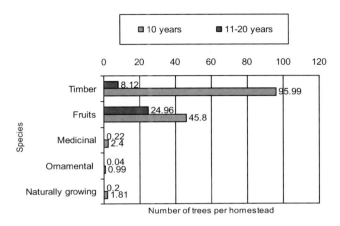

Figure-14. Existing tree species per homestead in the studied areas of the southwestern coastal region of Bangladesh.

5.10 Plant biodiversity protection and regeneration

Seeds, seedlings and other planting materials are the sources for improvement of plants and gene pools as well as regeneration and conservation of genetic base. Homestead is serving traditionally as a store house of seeds and seedlings as planting materials. The findings in relation to those are described below.

5.10.1 Types of species planted during last 10 years

Plantation is one of the major initiatives of plant protection and regeneration. Intensive tree plantation in the homesteads during the last 10 years gained momentum as shown in Appendix-6. In Table-27, it could be seen that during the last 10 years, farmers had planted 91.88 number of plants per homestead of which timber-yielding and others were 65.88 (71.70%) and fruit-yielding plants were 26.0 (28.30%). On an average, different seedlings (fruit, timber and others) planted by different farm categories per homestead were as follows: landless 45.46 in number, small 48.88, medium 125.59 and large 122.85. Specific species planted, were 23.30, 22.33, 17.65, 0.48, 0.40, 4.23, 4.01, 2.73, 2.13, 1.14 and 1.09 for Mahogony, Chambol, Raintree,

Bamboo, Koroj, Coconut, Betel nut, Guava, Jackfruit, Palmyra palm, and Date palm, respectively, which showed a positive co-relation with the relative prevalence values in the study areas. Species planted in less number were Bamboo, Koroj, Eucalyptus, Indian almond, Devils tree, Black myrobalan, Devil's cotton, Gulmohur, Banyan tree, and some other indigenous fruit and medicinal species were Monkey jack, Custard apple, Indian dillenia and Elephant apple etc. However, massive plantation of specific three timber-yielding species (Chambol, Mahogony and Raintree) was not desirable, because these species may be damaged easily by any natural disaster or pest and diseases. It would also create problem for timber supply system and also homestead ecosystem. For example, cyclone Sidr 2007 destroyed Mahogony, Raintree and Chambol along with other species which decreased market value of these species.

Table-27. Tree plantation per homestead during last 10 years in the southwestern coastal region of Bangladesh.

Tree species	Number of tree species planted/homestead				Average
	Landless	Small	Medium	Large	
Fruit-yielding	15.63	25.92	29.97	34.93	26.0
	(34.38)	(30.54)	(23.86)	(28.43)	(28.30)
Timber-yielding and others	29.83	58.96	95.62	87.92	65.88
	(65.62)	(69.46)	(76.14)	(71.57)	(71.70)
Total	45.46	84.88	125.59	122.85	91.88

* Figures in the parentheses denote the percentages.

5.10.2 Sources of planting material

Irrespective of farm categories, respondents reported to have collected seedlings of timber, fruit, medicine and others from different sources (Table-28). It was found that rural markets were the best source of supplying planting materials for all farm categories. It was also noticeable that both GO and NGO sources of planting materials supply were not satisfactory. Self source and relatives are still playing an important role as sources of planting materials, especially for Coconut, Betel nut, Jujube, Palmyra palm, Date palm, Banana and Bamboo etc. Sultana (2004) found the same trend where village market and own production were the main sources of planting materials. Private entrepreneurs raised large number of seedlings and sold those in the rural markets. They produced large number of seedlings of high demanding species i.e. Mahogony, Raintee, and Chambol. They usually ignored to supply minor fruit-yielding and less common species and were less concerned about the conservation of plant biodiversity. Considering the production and quality of timber and fruit planting materials over time, a policy guideline for all the growers need to be developed with special attention to source of mother plants (elite germplasms) and quarantine system. Farmer's choice and plant diversity perspective shall have to be incorporated to promote quality of planting materials in the homesteads of Bangladesh.

Table-28. Sources of planting materials used by the respondents of the study areas of the southwestern coastal region of Bangladesh.

Sources of seedlings	Distribution of respondents regarding use of sources of planting materials			
	Landless	Small	Medium	Large
Self production	7 (14.58)	22 (21.57)	12 (20.00)	6 (20.0)
Village market	38 (79.17)	88 (86.27)	47 (78.33)	27 (90.0)
NGO Nursery	1 (2.08)	4 (3.92)	4 (6.67)	2 (6.67)
GO Nursery	2 (4.17)	6 (5.88)	10 (16.67)	00
Relative	7 (14.58)	11 (10.78)	13 (21.67)	4 (13.33)

* Figures in the parentheses denote the percentages of the respondents.

5.10.3 Traditional practices in producing seedlings

Some indigenous knowledge base on traditional methods of seedling production were documented for creating interest among different stakeholders and for further investigation.

Household seedling production: It was found that Coconut, Betel nut, Palmyra palm and Guava seedlings were grown under the shade, while Papaya were grown in-house. Through this process, farmers used to select the best quality planting materials of fruit for further production as a reliable and improved source of seeds. This was a selection process for improvement of genetic base.

Budding process: Some fruit-yielding species, especially Jujube in which bud-grafting was widely practisced and this practice was reported to improve production in moderately saline to strongly saline areas. This has increased homestead gene pools and increased production through tree improvement.

Earthen pots: Most of the seedlings found in the rural markets were supplied from Sarupkathi, Banaripara upazillas of Pirozpur district (an area which is predominantly developed for plant nursery as a big enterprise). Most of the seedlings were raised in earthen mini-pots which called *"Tali method"* in which the main root was damaged, coiled, or cut and adventitious or secondary roots were developed.

Floating seedbed: It is an indigenous method for vegetable seedling production. Using water hyacinth or bamboo-made baskets was a common method of raising vegetable seedlings on floating seed bed. This method is popularly known as *"Kandi Ber"* system. Seeds are usually placed in small pouches made by semi-decomposed grass, known as *Hygrorhiza arista* (Dal ghas). This needs to be improved through further investigation.

5.10.4 Types of seeds stored in the homesteads

Rural homesteads are considered as *in-situ* gene banks where farm families collect and preserve seeds using their own knowledge. This has been playing an immense role in crop production and agrobiodiversity conservation in Bangladesh. Findings presented in Figure-15 showed that 71.25% farmers collected and preserved vegetable seeds, followed by 33.33% pulse seeds and 14.17% other seeds (Chilli, Watermelon etc.) in the homesteads. It implied that majority of the households preserved vegetable seeds for their own cultivation. It is noteworthy to promote homestead vegetable cultivation and indigenous species conservation. On the other hand, cyclone Sidr of 15 November 2007 damaged a major bulk of preserved vegetable seeds in the study areas. During post Sidr time, they had suffered seriously for sowing winter vegetables. Therefore, it was felt that community seed bank should be developed to tackle this type of natural disasters. However, it was found that some indigenous species like Chilli, Sesame, Mustard, Groundnut, Watermelon and Melon seeds were decreasing. They didn't preserve them in the homesteads. This showed a decreasing trend in agricultural biodiversity stock from this region. Pulse seeds, such as Bush bean, Khesari, and Mung bean were grown well but seeds of these species as propagules were decreasing from this region. However, the southwestern zone is still rich in indigenous and local agricultural cultivars, but they should be properly preserved.

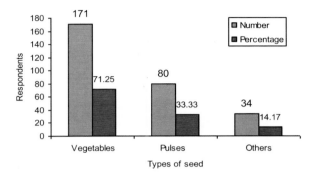

Figure-15. Types of seed collected and stored in the homesteads of the studied areas of the southwestern coastal region of Bangladesh.

5.10.5 Indigenous seed preservation/storage techniques

Farm families preserved/stored the cultivars using local and indigenous techniques (Table-29). It was found that use of plastic pot/basket/container was the highest (57.50%) technique followed by earthen pot (55.42%) known as "Motki" and bamboo-made indigenous basket (Dola/Shazi) (52.08%). These were used for preserving paddy and pulse seeds. Vegetable seeds were found to be preserved (52.08%) in glass pot or bottle or plastic jug replacing local earthen pot or open space. This was a sign of changing towards adoption of improved techniques. On the other hand, demand for big-sized plastic or iron-made container (drum/kati) used (7.92) for pulse, chilli and others were increasing. Homestead seeds were considered as the traditional source for preservation and conservation of agro-biodiversity. Women played the dominant role in processing, storing and maintaining them in the homesteads. According to Wakley and Momsen (2007), seed management was seen as an extension of women's domestic duties. It was learnt that framers were not familiar with the knowledge of modern preservation and storage techniques. So there was a need to promote better techniques for conservation and maintenance of agribiodiversity of this region.

Table-29. Indigenous seed preservation/storage techniques in the homesteads of the studied southwestern coastal region of Bangladesh.

Storage preservation/Storage technique	Respondents opinion		
	Number	Percentage	Priority
Plastic pot with polythene	138	57.50	1
Earthen pot (big-sized motki)	133	55.42	2
Bamboo made basket (dola/shaji)	125	52.08	3
Glass pot (bottle/jar/jug)	125	52.08	4
Earthen pot (small-sized pot)	57	23.75	5
Drum/Kati (container)	19	7.92	6

5.11 Felling of trees from the homesteads

Tree is considered as an income generating item including diversified utilization for household's uses. Tree felling is also common for mitigating the increasing demand of timber and fuel. There is a relation between need (income) and felling of trees which affects negatively the diversity and ecosystem of the homestead. However, it was hard to estimate different type of species felled from the homesteads over time. During enumeration, the whole homestead was visited to identify the places from where trees were felled. Besides, recalling from their memory was also helpful to note the felling trend of trees during the last 10 years.

5.11.1 Felling trend of trees during the last 10 years

Felling of fruit-yielding and timber-yielding trees of different farm categories was noted in Appendix-8. The most common species felled from the homesteads were presented in Table-30. It was found that both timber-and fruit-yielding species were felled during the last 10 years. In total, 37 different types of major species of different categories were felled of which timber-yielding and fruit-yielding trees were dominated. The most common species felled during the last 10 years were Raintree, Chambol and Mahogony.

72

Table-30. Dominant tree species felled in the homesteads during the last 10 years at different farm categories in the study areas of the southwestern coastal region of Bangladesh.

Species name	Respondent opinion regarding trees felled/homestead during last 10 years			
	Landless	Small	Medium	Large
Timber-yielding species				
Raintree	11(2.67)	24 (4.98)	16 (4.83)	15 (4.30)
Chambol	15 (1.25)	20 (1.19)	7 (1.25)	10 (2.13)
Mahogony	10 (0.96)	20 (1.27)	6 (0.80)	15 (2.83)
Babla	5 (0.19)	4 (0.10)	7 (0.23)	10 (0.47)
Jilapi	10 (0.77)	10 (0.70)	10 (0.60)	5 (0.60)
Fruit-yielding species				
Aam	10 (0.58)	10 (0.490	5 (0.23)	10 (0.93)
Tal	2 (0.23)	11 (0.57)	7 (0.77)	6 (0.67)
Khejur	5 (0.33)	20 (0.74)	13 (0.92)	5 (0.57)
Narikel	2 (0.08)	8 (0.29)	6 (0.28)	7 (0.63)
Supari	5 (0.13)	10 (0.20)	2 (0.03)	9 (0.47)

* Figures in the parentheses denote the percentages of the respondents.

5.11.2 Felling trend of trees based on age groups (maturity) of trees

Felling trend of trees of different age groups is shown in Table-31. Majority (80.93%) of trees irrespective of farm categories were felled which belonged to age limit ranging from minimum 10 years to maximum 20 years. Landless farmers felled trees in the age group 10 to 20 years (87.19%). It gradually decreased with a positive relation of small (85.01%), medium (77.12%) and large (76.99%) farm categories. This indicated that marginal groups (landless, small) of farmers had cut their trees relatively in short gestation period due to many socio-economic consequences and for meeting emergency needs of the family. Contrarily, large farmers kept/retained old trees for 21 to 30 years (20.29%) and above 30 years (2.72%) and felled them in long gestation period which was not eco-friendly to increase homestead production and for proper utilization of the micro-sites of the homesteads. On the other hand, on an average, landless farmers felled about 9 trees/homestead of different ages which gradually increased for small (12.17 trees/homestead), medium (11.58 trees/homesteads) and large (15.93 trees/homesteads). There was a positive correlation between felling and farm size, where average felling trend increased with the increase of farm size. However, increasing awareness among the farmers to fell trees in proper age would improve homestead productivity and proper space utilization as well as homestead biodiversity.

Table-31. Felling trend of trees based on age group of trees of the studied areas of the southwestern coastal region of Bangladesh.

Farm category	Trend of trees felled based on age group			Average plants felled/ homestead
	<10-20< years	21-30 years	>30 years	
	Plants/ homestead	Plants/ homestead	Plants/ homestead	
Landless	7.94 (87.19)	1.02 (11.21)	0.15 (1.60)	9.10 (100%)
Small	10.34 (85.01)	1.39 (11.44)	0.39 (3.22)	12.17 (100%)
Medium	8.93 (77.12)	2.38 (20.58)	0.27 (2.30)	11.58 (100%)
Large	12.27 (76.99)	3.23 (20.29)	0.43 (2.72)	15.93 (100%)
Total	39.48 (80.93)	8.02 (16.45)	1.25 (2.56)	48.78 (100%)

* Figures in the parentheses denote the percentages.

5.12 Management practices of homestead production system

The management of plant resources, especially trees and vegetables production from seed to seed are important for increasing production. Appropriate utilization of inputs and local resources, use of appropriate technology for ensuring maximum production are some dimensions of production and management. Identification of the management system practised by the households in cultivating plants in the homesteads have been discussed accordingly.

5.12.1 Management practices

Proper management practices of the homesteads can augment production and productivity as well as maintain vegetation properly. Six common management techniques and practices were found in this region as shown in Table-32. Farmers were reported to use local and indigenous practices for management of homestead production. It was found that above 56.66% of the farmers regardless of farm categories used organic manure, such as cow dung and compost etc. for vegetable and fruit production. Whereas use of chemical fertilizers was less in comparison to organic manure which was 19.5%. Farmers' tendency towards using chemical fertilizers for increasing homestead production was less predominant except large farmers. Usually, the large farmers buy chemical fertilizers for their field crops, but some of them also used for homestead vegetable production. Another aspect of management was earthing-up practiced by (49%) farmers of all categories. This earthing-up during dry season is done for Coconut, Lemon, Guava, Jujube and Mango and Mahogony etc. Majority of the respondents practised thinning and pruning along with weeding for the management of homestead production, and this was practised as follows: landless 47.92%, small 50.0 %, middle 61.67% and large 66.67%. In general, Coconut and Palm trees are managed by cleaning head and cutting leaves. Salt is used for protecting insects, especially ants and to increase production of Coconut. It was also found that 4 to 13% of farmers used fencing for vegetable or fruit production but at the same time for providing less effort to protect vegetation of the homesteads. Apart from these practices, spraying irrigation, even in dry season was almost absent. Therefore, scientific management practices along with the use of surface water is needed for commercial agro-sylviculture in this region.

Table-32. Management practices followed by the respondents in the homestead production system of the studied areas of the southwestern coastal region of Bangladesh.

Management practices	Farm category			
	Landless	Small	Medium	Large
Organic manure	27 (56.25)	61 (59.80)	34 (56.67)	14 (46.67)
Chemical fertilizers	6 (12.50)	18 (17.65)	12 (20.0)	11 (36.67)
Earthing-up	23 (47.92)	51 (50.00)	29 (48.33)	16 (53.33)
Thinning	5 (10.42)	5 (4.90)	6 (10.0)	5 (16.67)
Pruning	18 (37.50)	46 (45.10)	31 (51.67)	12 (40.0)
Fencing	5 (10.42)	13 (12.75)	12 (20.0)	4 (13.33)

* Figures in the parentheses denote the percentages.

5.12.2 Sorjan-an indigenous technique used for the regeneration of biodiversity

'Sorjan' is an indigenous method, especially used by the farmers of Southern zone for vegetable and fruit production. In Indonesia, it is a common crop production technique in the coastal areas. 'Sorjan' is a Javanese (Indonesian) word which means "two" or "graft". Two or more trench is prepared between two rows along with the raised beds. This is why, Sorjan is popularly known as 'Kandi Ber' (raised bed) in southern zone of Bangladesh. Lands of these areas are low-lying and are affected with salinity during the monsoon. As a result, farmers can produce crops only once in a year but side by side, 'Sorjan' method can produce higher quantities of fruits and vegetables in the lean season. Therefore, it is considered as a fruitful technique for increasing farm production and income.

Suitable time for the preparation of 'Sorjan' is during January to February and planting time is March to April at the beginning of monsoon. Both low value and high value vegetables are grown in this method. The low value vegetables were: Amaranths, Country bean, Spinach, Arum, Chilli, Sweet gourd and high price: Brinjal, Snake gourd, Bitter gourd, Bottle gourd and Ladies finger. However, Brinjal and Tomato perform well in 'Sorjan' model. Suitable trees such as: Kapila/Jiga *(Lannea coromandelica)*, Mandar (*Erythrina veriegata*), Dhaincha (*Sesbania sesban*) are commonly grown in Sorjan. These plants are also suitable for using them as hedge and live fence and to support climbing vegetables like beans and cucurbits. 'Sorjan' has become a potential technique for small size fruit and vegetable production as well as regeneration of plants biodiversity in the marshy and wetlands of the homesteads. However, production technology and profitability are still unascertained. Field level information on these aspects would be helpful to create awareness among the farmers.

5.13 Role of gender/women in homestead plant biodiversity management and conservation

Women's role in the management including plantation, nursing, protection, decision-making and use of homestead resources have been shown in Table-33. Majority of the respondents (34.58%) sought women consultation for the well-being and overall progress of the family. Women play an intensive role in the management of homegarden. They also played vital role in the decision-making process (15.83%). Women were found to spend more time, especially in the agro-sylvicultural management, nursing and intensive labor for plant protection and survival in the homesteads (12.50%). Side by side, women's role in financial management (11.67%), their equal right (10%), decision-making for selling trees (4.54%) were also recorded. Gender perception of men and women is changing in a positive way and awareness has been increasing among the people, may be this is the consequence of the private-public publicity through mass media and motivation also. Contrarily, a proportion of the respondents were reluctant to accept women's role with the notion that women need not know more in the management (5.42%), followed by no right and access (3.33%) over the homestead resources. These reflect a common arrogance and negative attitude toward women's role over the homestead production and management.

Women are taking predominating initiative in plant biodiversity management as housewives, home gardeners, herbalists, seed custodians, and informal plant breeders. In most cases, plant use, management and conservation occur within the domestic realm where they are largely "invisible to outsiders" and are easily undervalued. CBD recognizes "the vital role that women play in the conservation and sustainable use of biological diversity" and affirm "the need for the full participation of women at all levels of policy-making and implementation for biological diversity conservation. FAO, through its Commission on Genetic Resources for Food and Agriculture, should develop a gender code to facilitate the recognition of women's role in the conservation and enhancement of agro-biodiversity in the operational framework for Famers' Right as well as for contribution of women in the conservation and selection of valuable genetic diversity in the home gardens. Prioritizing the conservation of plants, women should be the curators for promoting traditional knowledge and practices; recognizing indigenous rights. Women's rights and access over the plant resources need to be conserved. Ensuring women's full participation in decision and policy formulation is essential for homestead plant biodiversity conservation and regeneration.

Table-33.Gender/Women role in homestead biodiversity conservation and overall management.

Women's role	Respondents' opinion regarding women's role in biodiversity		
	Number	Percentage	Ranking
Consultation with women: need consultation with women for the benefit of family development, achieving success and managing the entire family work including tree and vegetable management.	83	34.58	1
Women's opinion in decision-making: women's (wife and mother) opinion is essential in the felling, plantation and management. Taking jointly family decision in this purpose gained force for overall establishment of the family.	38	15.83	2
Women spending longer time in the homesteads: Women spent more time in the preservation and conservation of homestead plants through rigorous nursing and labor.	30	12.50	3
Homesteads saving/utilization: Women are very active in providing family income received from selling of homestead products and use these money for paying credit and regular installment.	28	11.67	4
Women should have equal right: In taking decision in the household level they must ensure their right like the male.	24	10.00	5
Women select/identify the trees to be sold or felled and to be planted: Women can play a key role in this selection processes.	11	4.58	7
Women need not to understand: women were unfit to consult regarding homestead natural resources management.	13	5.42	6
No right and access over homesteads' resources: Women posses no authority over homesteads' resources. They are not entitled to get access over any resource of the homestead.	8	3.33	8

5.14 Problems faced by the farmers in homestead production and management system

The study documented the major problems faced by the farmers in homestead production and management system as shown in Table-34. A total of 12 major problems were identified including their priority relating to management and production system, knowledge and skill base, natural resources constraints, physical and social problems which were collected during the study and summarized as follows.

Table-34.Major problems faced by the farmers in homestead production and management system of the studied areas of the southwestern coastal region of Bangladesh.

Problem	Respondents opinion (%)	Ranking	Respondents' opinion regarding problems faced and nature of damage
Lack of advanced knowledge and technology	57.91	1	Shortage of advanced knowledge specifically: i) modern varieties, ii) cultivation and planting methods, iii) seed storage and seedlings raising. During natural disasters, farmers have had no proper knowledge for the protection and conservation of food and natural resources
Pest and diseases infestation	52.11	2	Fruits and vegetables are damaged by pests and diseases. Ten pests and diseases were identified for this region: i) Red pumpkin beetle, ii) Fruit fly, iii) Vein clearing disease (virus disease) of country beans, iv) Cut worm insects of vegetables, vi) Panama wilt of Banana, vi) Stem weevil of Banana, vii) Shoot and fruit borer of Brinjal, viii) Vein clearing disease of Okra, ix) Fruit dropping of cucurbit vegetables, x. Eaten leaves and moulds of fruits and vegetables
Salt water intrusion	52.50	3	Salt water intrusion and prolonged drought were the emerging and acute problems which were hampering overall components of homestead production, such as death or damages of plants, vegetables, fish, poultry and livestock
Improper homestead space planning and utilization	49.58	4	Haphazard and irregular spatial utilization of homesteads has been reducing homesteads' production, especially in age-old homesteads. Over-aged plant canopy have covered the whole homesteads which were resulting long time shade hampering vegetable production in most of the homesteads. It was noticeable that farmers were unaware about the rotational time for felling the trees. In some homesteads, huge number of big-sized plants existed which reduced overall homestead productivity
Lack of maintenance of embankments and sluice	40.83	5	Faulty sluice-gate and lacking of proper management of embankments have affected water management in the study areas. These problems created diversified problems which an ultimately hampering homesteads as well

gates			as agricultural production in this region
Problem	Respond ents opinion (%)	Ran king	Respondents' opinion regarding problems faced and nature of damage
Low productivity of homesteads	40.0	6	Homestead production of vegetables and fruits has been decreasing in this region compared to the past. Low yield of economic and profitable fruit-yielding species such as Coconut, Tal, Supari and Banana and other species were notable. The number of this populations has also been decreasing from the homesteads gradually
High cost of inputs, seeds/seedlin gs	36.66	7	High price of seeds and seedlings along with an increasing trend of fertilizer and pesticide use have created a negative impact on growing vegetables and rearing plants in the homesteads
High labor cost and shortage of labors. Migration to city areas,	28.75	8	Shortage of agricultural labor and high wage also a big problem for these areas. Homestead owners complained against the wage labouers. They are migrating to cities or towns in the dry season while it was the peak time for managing the homesteads (pruning, thinning of plants, fence preparation and other management practices)
Recurring natural disasters	28.33	9	Super cyclone Sidr 2007 destroyed the natural resources of homesteads, especially diversity of plants. About 80-90% homesteads and their houses were damaged with plants, poultry, livestock and other lives in the homesteads
Canals siltation	22.91	10	Most of the canals of this region have silted-up which were unable to flow water in dry season and drainage during rainy season
Land scarcity	20.41	11	Scarcity of land is a common problem to accommodate more households in the same homestead. The problem affects especially the marginal farmers and their income
Social problems	15. 83	12	Clash among and between the households of the homestead which affect the overall management and production of the homestead.

5.15 Income and expenditure of the household from different sectors and sub-sectors

For assessing the income and expenditure of the households, special attention was paid to their experience, recalling from their memory, tentative production and product cost and sale value of the products. This made it easier to estimate the total income and expenditure of a family.

5.15.1 Household income

The study assessed income generation by the household of a family as shown in Figure-16. Three major categories of income were identified and these were: i) field crops, ii) homestead and, iii) others. The farm income (homestead and field crops) contributed to 69.49% of total income. Between the income of two sectors, contribution of field crop was 40% and contribution of homestead was 30% on an for average of all the farm categories. The remaining 30% of the income were obtained from other sources, which included off-farm activities such as business, handicrafts, salary and labor wage etc. Alam *et al.* (1995) also found about 64.21% of cash inflow into small household were from crop sector including the sale of homestead products. Therefore, one third of the total income was earned directly from the homestead. This can be looked upon as **"reserve bank"** of "food and cash" compared to less investment and inputs. Besides, many other indirect contributions of homesteads were playing significant role as life supporting system of the farmers. Analyzing the homestead income (apart from field crop and other income) it was found that landless contribution was higher (31.77%) than small (33.02%), medium (27.47%) and large (29.19%) farmers where average contribution was 29.49%. The reason for less contribution by large households was that large farmers highly depended on field crop production and gave less attention to home gardening. On the contrary, landless and small households were involved more intensely in their homestead and getting relatively more income. Therefore, homestead as a source of diversified natural resource base can contribute more in the total income of a rural family.

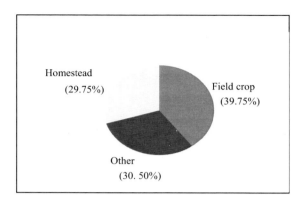

Figure-16. Household income of the respondents from different sources (expressed in %) of the studied southwestern coastal region of Bangladesh.

5.15.2 Contribution of homestead income from different sub-sectors

The contribution from different sub-sectors (timber, fruit, vegetables, fisheries, livestock and poultry) to homestead income by the different farm categories (Figure-17) was further assessed. It was found that contribution of timber selling in total homestead income was the highest (28.88%), followed by livestock (19.31%), fruit-selling (15.54%), fishery (15.17%), poultry (10.90%), and vegetable (10.20%). It was clear that homestead income was a cross-cutting issue where the household depended on cumulative contribution of the various sub-sectors of natural resources. Contribution of fruits and vegetables to the total homestead income was not satisfactory. This was due to prolonged salinity and drought in this region. Considering the land size of the homesteads of this region, fruit and vegetable production should be extended in the light of commercialized approach. Reduction of poverty through increasing homestead production and daily income need to be paid especial attention for the marginal farmers.

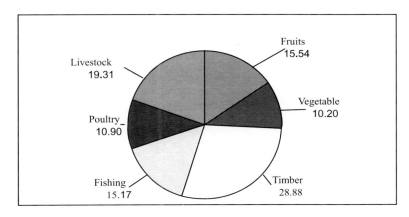

Figure-17. Contribution of homestead income from different sub-sectors by all the farm categories of the studied southwestern coastal region of Bangladesh.

5.15.3 Household expenditure

The study assessed household's total expenditure (person/year) of a family of different farm categories (Table-35). The annual expenses of different farm categories were as follows :landless (tk.11, 577.39), small (tk.15, 739.01), medium (tk.22, 269.48) and large farmers (tk. 29,686.33). It was observed that landless farmers spent below one thousand taka per month including all the items where large farmers spent about three thousand taka per month. This revealed a miserable living standard of the rural poor of these areas.

Table-35. Household expenditures according to farm categories of the studied southwestern coastal region of Bangladesh.

Items of expense	Expenditure (Taka/person/year)			
	Landless	Small	Medium	Large
Education	990.90	1448.73	1939.38	2821.84
Daily expenses	2613.15	3386.86	4716.56	5912.76
Rice	4466.88	5200.46	7474.66	7349.01
Fuel	492.33	494.79	595.46	765.26
Edible oil	685.05	889.24	1044.95	1133.45
Clothes	601.94	815.20	1110.19	2135.06
Medicare	635.92	1020.39	1190.92	1867.82
Festival	725.12	851.84	1407.00	1850.57
Agriculture	248.79	1163.63	2271.64	4511.49
Others	117.31	467.87	518.72	1339.08
Total	11577.39	15739.01	22269.48	29686.33

5.15.4 Contribution of different items of household expenditure and their distribution

Contribution of different items of expenses and their distribution are shown in Figure-18. The relative share of individual component of expenditure on rice was the highest (32.07%) out of total cash outlay. The second item was daily or periodic expenses which included fish consumption, vegetable and spices purchase, hospitality, transport cost and others (21.19%), followed by agricultural production (9.32%), education (9.04%), medicare (5.98%), festivals (5.97%), clothes (5.56%), edible oil (4.99%), fuel (3.01%) and others (2.87%). It was found that for fulfilment of the basic needs of landless and small households, majority of their income was spent on food, daily necessities, clothes and medicine. On the other hand, more than one third income was incurred only for rice which was hard for the poor who purchased it round the year.

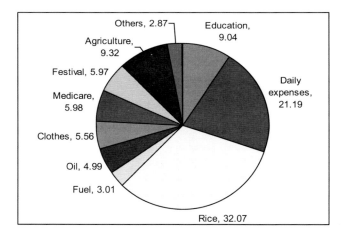

Figure-18. Proportion of different items of household expenditure in the homesteads of the study areas of the southwestern coastal region of Bangladesh.

5.16 Food security status (food over time, food habit and storage) of the households

For assessing food security, three indicators were set which characterized the food status (food over time, food intake habit and food storage) of the farmers. Likewise, three simple criteria were set-up to assess the food security during the last 10 to 20 years (good, moderate, poor and fragile).

5.16.1 Food security over time

Food security status of the household between 10 to 20 years ago and present time was compared and presented in Table-36. The overall food security of the respondents had improved where most of the fragile and poor farmers in the past turned into least poor and moderate poor categories. It was found that 71.4% of landless farmers remained fragile in the past, but at present they have turned themselves into poor and moderate category which implied their improvement in ability to purchase food. On the contrary, 3% of the landless farmers remained in good condition in the past but the situation, however, deteriorated to 1.3%. These also proved that food security to some of the farmers decreased over time which was due to mainly socio-economic consequences such as natural disasters, loss of earning members and sources etc.

Table-36. Food security of the respondents over time in southwestern coastal region of Bangladesh.

Farm category	Food security of the respondents							
	10-20 years ago				Present time			
	Good	Moderate	Poor	Fragile	Good	Moderate	Poor	Fragile
Landless	3 (7.0)	13 (11.2)	22 (32.8)	10 (71.4)	1 (1.3)	22 (18.2)	23 (60.5)	2 (100)
Small	13 (30)	54 (46.6)	33 (49.3)	2 (14.3)	26 (32.9)	64 (52.9)	12 (31.6)	00 --
Medium	9 (20)	38 (32.8)	11 (16.4)	2 (14.3)	34 --	23 (19.0)	3 (7.9)	00 --
Large	18 (41)	11 (9.5)	1 (1.5)	00	18 (22.8)	12 (9.9)	00	00 --
Total	43 (100)	116 (100)	67 (100)	14 (100)	79 (100)	121 (100)	38 (100)	2 (100)

* Figures in the parentheses denote the percentages of the respondents.

5.16.2 Food intake by the households

Food intake and cooking habit by the households were also changing as depicted in Table-37. It was observed that 89.2% of the respondents have had taken three meals a day and 10.8% two meals a day, this is a good sign of changing food habit. In general, average quantity of food consumed (apart from calories consumption) in Bangladesh was increased from 886 grams in 1991/92 to 948 grams in 2005, an increase of 7%

over time (Mandal, 2008). The common pattern of food habit of having water-soaked rice in the morning and evening has been replaced gradually. However, about 37.9 % of the marginal farmers still have had two meals a day. They generally cook in the morning (8.30 to 10 am.) and keep it for noon and daily labor class cook their food late at night for supper.

Table-37. Food intake by the households of the studied areas of the southwestern coastal region of Bangladesh.

Farm category	Food intake		Total
	Three meals/day	Two meals/day	
Landless	35 (72.9)	13 (27.1)	48 (100)
Small	91 (89.2)	11 (10.8)	102 (100)
Medium	58 (96.7)	2 (3.3)	60 (100)
Large	30 (100)	---	30 (100)
Total	214 (89.2)	26 (10.8)	240 (100)

* Figures in the parentheses denote the percentages of the respondents.

5.16.3 Stored food in the homesteads

Food storage is another parameter of food security in the rural areas (Table-38). About 48% of the respondents were not involved in food storage and 52% of the respondents were involved in subsistence food storage during the rainy season. Majority of the farmers of large and medium households were engaged with at least one item for food storage. The major stored items were: i) rice, and ii) pulse, while the minor items were: i) chilli, ii) sweet potato and iii) molasses.

Table-38. Stored food in the homesteads by the different farm categories of southwestern coastal region of Bangladesh.

Farm category	Respondent distribution regarding food storage	
	Stored	Not stored
Landless	04 (3.20)	44 (38.26)
Small	54 (43.20)	48 (41.73)
Medium	37 (29.60)	23 (20.0)
Large	28 (22.40)	02 (1.73)
Total	125 (52.08)	115 (47.92)

* Figures in the parentheses denote the percentages of the respondents.

5.17 Relative contribution of the homestead tree and farm products

Homestead tree contributes as a source of income while vegetables help nutrition and livelihood support. On the other hand, on-farm and off-farm activities of the households were also an important source for food security and income generation, this has been discussed in this section.

5.17.1 Contribution of homestead trees product

The contribution of homestead trees product is important to understand its significance in the livelihood support as shown in Table-39. It was found that trees contributed in diverse ways in the life supporting system. This is summarized into 7 events. The major contribution of trees was food, fruits and vegetables (47.50%). The second highest contribution was cash money (47.08%), followed by risk coverage (35.42), timber and fuel wood (27.08%), organic fertilizer (16.67%), and furniture/utensils (13.75%). Besides, farmers preferred trees for meeting many of their seen and unseen needs of daily life. It was concluded that homestead trees contributed as a source of "food and cash" and chain of energy in the ecosystem for all the living organisms.

Table-39. Contribution of homesteads trees for livelihood support of the respondents of the studied areas of the southwestern coastal region of Bangladesh.

Events of contribution	Distribution of respondents regarding contribution of homestead tree product		
	Number	Percent	Ranking
Food, Fruits, Vegetables	114	47.50	1
Cash money	113	47.08	2
Risk coverage	85	35.42	3
Timber wood and fuel wood	65	27.08	4
Wooden made construction and materials	60	25.00	5
Organic fertilizer, leaf and shade	40	16.67	6
Furniture /Utensils	33	13.75	7

5.17.2 Contribution of homestead vegetables

The study assessed contribution of homestead's cultivated and naturally growing vegetables as mentioned in Table-40. Contribution of homestead vegetables was the highest (55.59%), followed by naturally growing vegetables (15.66%) and purchased vegetables (28.75%). The contribution of homestead cultivated vegetables was maximum for medium farm category (58.25%), followed by small (57.54%) and landless (54.27%) and large (52.3%) farm categories. The contribution of naturally growing vegetables was maximum for landless farm category (22.65%), which gradually decreased as the farm size increased. It was found that naturally growing vegetables have been supporting enormously, especially for the landless and small farmers in their food intake and nutrition. On the other hand, 37.37% of vegetables were purchased by the large farmers, but the trend decreased as the farm size decreased. However, it was noted that farmers of all categories purchased vegetables from the local markets. Obviously, it proved the scarcity and low productivity of vegetables in the southwestern coastal homesteads. Scarcity of vegetables caused malnutrition in this region. This is a common phenomenon in the socio-economic life of the people of this region (Novib, 1996). It was observed that dependency on naturally growing vegetables is increasing, especially during dry season (during high salinity and drought).

Table-40. Contribution of homestead's cultivated and naturally growing vegetables in the homesteads of studied areas of the southwestern coastal region of Bangladesh.

Farm category	Distribution of respondents regarding contribution of vegetables (%)			Total %
	Cultivated vegetables	Naturally-growing vegetables	Purchased vegetables	
Landless	54.27	22.65	23.08	100
Small	57.54	15.3	27.16	100
Medium	58.25	14.37	27.38	100
Large	52.3	10.33	37.37	100
Mean	55.59	15.66	28.75	100

5.17.3 Contribution of on-farm and off-farm activities

Relative contribution of on-farm and off-farm activities of the farm families is presented in Table-41. The role of farm activities for the livelihood dependencies by all the farm categories was 68.62%, of which landless was 52.60%, small 67.85%, medium 73.33% and large 84%. Therefore, farm activities are still a major contributor in the livelihood of the cross-section of the farmers. On the other hand, the dependencies on off-farm activities were 31.38% for all the farm categories where landless had 47.40%, and small farmers 32.15% implying that their dependencies for livelihood and earning were increasing.

Table-41.Contribution of On-farm and Off-farm activities of the southwestern coastal region of Bangladesh.

Farm category	Distribution of respondents regarding on-farm and off-farm activities			
	On-farm activities		Off-farm activities	
	Contribution %	Standard deviation	Contribution %	Standard deviation
Landless	52.60	39.37	47.40	39.37
Small	67.85	30.03	32.15	30.63
Medium	73.33	27.47	26.67	27.47
Large	84.00	19.18	16.00	19.18
Total	68.62	31.68	31.38	31.92

5.17.4 Important major and minor species for food security

Major and minor plant species that are important for food security of the farmers are shown in Table-42. Some species were used for food security providing fruits during the various seasons and some of the species were used as "life supporting species", especially to combat crisis period. These life supporting species important, especially for the poor and marginal farmers. Besides, some citrus and tamarind-yielding species also played an important role for the farmers. However, Ahmed (1997) reported that about 31 minor fruit species had reached the stage of near extinction from the homesteads of Bangladesh. The present study also revealed some life supporting species reaching the level of rare or threatened stage.

Table-42. Homestead species dominant with regard to food security of the study areas of the southwestern coastal region of Bangladesh.

Local/English name	Scientific name	Uses/Characteristics
Aam/Mango	*Mangifera indica*	Household seasonal nutrition, cash income , timber and fuel
Amra/Hogplum	*Spondias pinnata*	Seasonal income and economic species
Boroi/Jujube	*Ziziphus mauritiana*	Highly profitable in strongly saline areas
Bel/ Wood apple	*Aegle marmelos*	High price, wider medicinal uses
Beelati gab	*Diospyros blancoi*	Minor fruit, cheaf, fleshy, grown plenty
Aitakala/Banana	*Musa paradisiaca*	Cheap and nutritious food for round the year and also use as "famine food"
Dhekishak/Table fern	*Diplazium esculentum*	Poor collect from nature, recipe with shrimp
Babla/Jilapi	*Pithecellobium dulce*	Post Sidr, 2007 people took it
Kachu/Boi/Loti	*Colocasia esculenta*	Common recipe with coconut and shrimp
Kathal/Jackfruit	*Artocarpus heterophyllus*	A family used it jointly, social, value
Khejur/Date palm	*Phoenix sylvestris*	Juice, molasses and fuel wood
Lau/Bottle gourd	*Lagenaria siceraria*	Cheap and available winter vegetable
Mistialu/Sweet potato	*Ipomoea maxima*	Crisis period food and vegetables
Mistikumra/Pumpkin	*Cucurbita moschata*	Crisis period food and vegetables
Mula/Radish	*Raphanus sativus*	Cheaf and consume it heavy
Narikel/Coconut	*Cocos nucifera*	Highly economic and multipurpose species
Panikachu/Taro	*Colocasia esculenta*	Poor and pro-poor vegetables

	var. aquatilis	
Pepey	Carica papaya	Economic and medicinal value
Shapla/Water lily	Nymphaea pubescens	Pro-poor nutrition, income, crisis period food
Supari/Betel nut	Areca catechu	High economic and social value
Tal/Palmyra palm	Borassus flabellifer	Palm, juice for poor, multipurpose uses

5.17.5 Utilization and status of "life supporting species"

The identified "life supporting species" are very important for human life and these are much known in the studied region. Among them, utilization and status of some species are given below:

Banana/Atiakala (*Musa paradisica*): In the southern region, Atiakala (*Musa paradisica*) is grown as an important tasty and delicious food to be used both as fresh fruit, vegetables and medicinal purpose. The fleshy plants are also called "famine food". The pedancle of the spadix inflorescence known as "thor" is used as an indigenous vegetable, especially by the poor section during the crisis period. During lean season, green banana is also boiled and eaten as a dish. The terminal end of the inflorescence, termed as "*mocha*" is also used as vegetable, very popular food and delicacy in southwestern region. This species is used as an indigenous medicine and has immense value in the community for treating dysentery, constipation etc. One bunch of Banana is sold in the retail market worth Tk.100-250 (comprising 50-250 fingers). This species is unable to tolerate salinity; however, Sidr has seriously affected and damaged this fruit-yielding plant. Besides, Banana growth is hampered under Rain tree and Chambol trees which decrease its cultivation in the homesteads.

Water lily/Shapla (*Nymphaea pubescens, N. rubra*): Red and white types of water lily were found to grow in the monsoon in the ponds and marshy lands in the southwestern homesteads. These species grow naturally in the homesteads. They have ample economic value for family nutrition and income generation. Most of the poor families collect the fleshy peduncle, cook with coconut chips. This is a very tasty recepie of vegetable in this region. Rural children and youths collect them and sell in urban markets as a source of income and employment during lean monsoon at a rate of tk.10-15 for a bunch of water lily peduncles. The plant parts are also used as medicine, especially for *N. nouchali*; corms of *N. pubescens* are boiled or fried and eaten by the children as a good source of carbohydrates; seeds are husked and cooked during famine or puffed ('*Dhaper khoi*') to make village candi ('*Moa*').

Hot chilli/Bombai marich (*Capsicum frutescens* 'Bombiai'): Grows in almost all the homesteads in this region where Chilli is commonly consumed. This Chilli is comparatively cheap which is especially preferred by the "poor family" and is also used during the crisis period. Chilli, made into pickles has become popular in urban areas. The population of this species has been decreasing due to heavy rain and natural disasters.

Water Primerose/Helencha and Haicha (*Enhydra fluctuans, Alternanthera philoxeroides*): These species are commonly eaten by all classes of people especially by the poor almost round the year. It is grown in the ponds, adjoining areas of the homesteads and also in the peripheral areas. The poor families cook it mixing with hot Chilli. During the dry and rainy season and post natural disasters, it is used as "life supporting leafy vegetables". These species can be popularized in the city areas through packaging and processing.

Yam/Kachu (*Colocasia esculenta*): Green kachu stem and leaves are popular and consumed round the year. The stolon known as "*Boi*" is a common recipe with crushed coconut and small shrimp. It has wider economic value in nutrition and income generation for selling in the rural and urban markets. A good number of seasonal unemployed youths are used to involve themselves to collect '*Boi*' and sell in the market. It is a ready source of income and employment during the late monsoon at a rate of tk.10-25 per kilogram. It is a rich source of vitamin-A, iron and other mineral elements.

Mamakala/Baoali lata: (*Sarcolobus carinatus*): This is used as a popular wild fruit and eaten by the farmers along with the children and women of this region. The green fruit is also used in making vegetable curry during the crisis period by the people of these areas.

Jilapi (*Pithecellobium dulce*): In southwestern zone, it provides nutrition to rural children during the months of May to June. During post cyclone Sidr, 2007 children and women took it as fruit. In the Philippines, this plant is also used as a major fruit tree.

Choila (*Sonneratia caseolaris*): This is a wild fruit-yielding species. Its green fruits are used as soup or curry, especially by the poor people. Further investigation may be done to know the status and mineral content of this fruit.

5.17.6 Status of the predominantly grown species in the homesteads

Indian buch/Karanja (*Pongamia pinnata*): It is a big tree grown in the homesteads for diversified uses (fuel, wood, timber, shade, agricultural tools, cottage industry and wind breaker). It has wider adaptability to salinity and grows well in marshy ecosystem around the homesteads. This species is also decreasing from the homesteads and adjoining areas.

Indian almond/Katbadam (*Terminalia catappa*): It is a big tree grown in the homesteads for diversified uses (fuel wood, timber, shade, agricultural tools and handicrafts) and adaptable to less to moderately saline and marshy ecosystem of the homesteads. Fruit (seed) is a favorite nut eaten by the children. The species is decreasing in number due to felling and natural disasters.

Coral tree/Mandar (*Erythrina veriegata*): It is a popular species for wider uses in the homesteads such as fuel wood, fence, fodder, shade for fruit orchards, especially betel nuts, coconut and other agroforestry purposes. The species is also decreasing in the homesteads for heavy felling.

Jiga/Kapila (*Lannea coromandelica*): This plant is used for making live-fence in almost all the homesteads in southwestern coastal zone of Bangladesh. Climbing vegetables like country bean, cucumber, and gourd grow well on this tree. Generally, branches of this tree are used for vegetative propagation. This species is also decreasing from strongly to moderately saline areas.

River ebony/Pechi gab (*Diospyros malabarica*): This is a less common and rare species; its ripe fruit is eaten by the children. Unripe fruits bear high tannin content which is used for boat, net and dyeing. This species has been degraded for massive felling for multipurpose uses. The gene pool of this species can be preserved for further study.

Hurmai/Urmai (*Sapium indicum*): Its fruit and leaves are used as botanical pesticide for seed preservation, storehouse of rice, paddy and pulses, fish staffing etc. *Sapium indicum* has been decreasing tremendously from the homesteads which needs to be conserved.

Indian oak/Hijal (*Barringtonia acutangula*): This species belongs to the homesteads of marshy and saline land. It is decreasing from the homesteads.

Bulrush/Hoglapata *(Nypa fruticans)*: It is a common palm species, naturally grown in the wetlands and ponds of the homesteads. Hoglapata is widely used by the people of the southwestern zone of the country for making dwelling houses, mats especially for paddy drying, fish packing and fuel etc. Many households use it for molasses (golpata "*Gur*") production and they dry its inflorescence for making flour for preparing indigenous bread. Village women prepare mats and earn cash money. This species can be grown successfully in the less saline and moderately saline zones.

Coriander leaf/Bilati Dhainapata (*Eryngium foetidum*): It is a common spice-yielding species in the homesteads. It has economic value in the city areas. The spice production is decreasing due to floods and droughts. The species needs to be preserved in the gene bank.

Polau pata (*Pandanus odorus*): It is a common spice-yielding species used for cooking hotchpotch (*paish/sirni*) and other indigenous item of foods in this region. It grows in the nooks and corners of the homesteads with less care. The gene pool of this species can be preserved.

Dalghas (*Hygroryza aristata*): It is an economically important fodder species grown in the ponds and adjoining swampy areas of the homesteads. The farmers, who have buffaloes in their farm, cultivate it during the off-season for fodder. It is also used in making pouches for raising vegetable saplings in '*Kandiber*' floating seed bed system. This species is decreasing in the swampy lands of the homesteads.

Water hyacinth/Kachuripana (*Eichhornia crassipes*): It is a non saline tolerant species grown in some of the ponds and ditches of the homesteads. It has wider use potential especially in making organic manure in the dry and winter seasons. It is used as raw materials for vegetable seedling raising, especially Bottle gourd, Cucumbers, Bitter gourd and Chilli in winter and as mulching materials for Coconut saplings in the homesteads and mulch for Potato, Water melon in the fields.

5.17.7 Indigenous knowledge base for homestead plant utilization

Indigenous knowledge base for homestead plant utilization is believed and practised by the farmers of these region's (Table-43). There were many available indigenous knowledge, folklores and proverbs which the farmers have learnt from their ancestors or neighbours. Some of them believe that, Tulsi/basil (*Ocimum tenuiflorum*) and Kalkephul/Lucky nut (*Thevetia peruviana*) are good for Hindu homesteads (Hindu Bari). They have learnt that by growing Country Neem (*Azadirachta indica*) in the Southern side purifies air and controls small pox. Apang (*Achyranthes aspera*) is commonly grown and its ring is used to control Jaundice (haldi palong). Woody trees in the back-yard protect house from theft and also keep the houses cool. Fruit trees nearby the homesteads ensure proper management and protection. There are some messages about the homesteads reported by Ms. Khana. This is called 'Khanar bachan'. It is commonly used and is popular in this region.

Table 43. Indigenous knowledge base on homestead plant utilization.

Local knowledge and proverbs	Meaning and utilization
Dakhin duari gharer raja, purbo duari tahar raja, paschim duarir mukhe chai and uttar duarir kichue nai, 'Khanar Bachan'	South and east-facing homesteads are considered to be superior and west and north facing ones are the worst
Dhakhine fak, uttare bag, purbe hash and pachim-e bash	Production technology of homesteads is as follows: South face for open space, north face for orchard and garden, east face for pond fisheries and duck raising and west face for bamboo as live fence and wind break
Narikele nun-mati, tara-tari badhe guti	Applying salt at the base of Coconut trees ensures rapid formation of inflorescences and large number of fruits
Bagan theke ashlo buri-sara gaeye tar kuri-kuri	Jackfruits from the home garden look like an old women with their spiny skin
Ekhta-bade satti bati, je na balte parbe tar nakti kati	A fruit bearing 7 persistent calyx lobes like a bowl is called Indian dillenia (Chalta)
Samayae na dai chash, tar dhukha baro mash	The farmers who do not till the land in time, suffer all the year round
Kacha-kale tuk-tuk pakle sindur, je na kaite parbe, harr nam thumu baitta indur	The fruit looks very green and beautiful in raw stage, but in the ripened stage, it looks to be very red (Mama sindur gach)
Chiri-chiri pata-tar sonali baran, pakile thoie, majile khai	The leaf of date-palm is feather like and its fruits are harvestd in mature stage and eaten after being fully ripend
Basat-bari anginai, sabji -chashee pusti jogai	Vegetable cultivation in home-yard, provides nutrients
Chari-dikee kata-kota maddha-khane jammadar-beta	Pine apple is borne at the middle of the plant surrounded by spiny leaves

5.18 Impact of homestead plant resources on income generation and livelihood support

Homestead plant resources play diversified roles in supporting or supplying major food is grouped into eight components along with their uses as depicted in Box-1. These are i) food (different vegetables, fruits and juice/molasses), ii) cash money (for family education, family maintenance and expenditure and repayment of loan, iii) safety net (risk coverage, have no option of land), iv) timber (house construction, boat making, shade making, shop/ agricultural tools, v) fuel wood, leaf and biomass, bamboo, vi) furniture (household utensils, reading materials, bed-stead), vii) environmental quality (wind protection, flood protection, beautification and organic matter), viii) social festival (marriage, dowry, gift, road, bridge and pole).

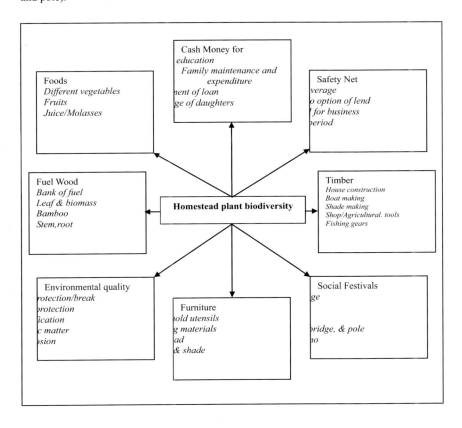

5.19 Conservation of homestead plant biodiversity

Bangladesh has signed and ratified the Convention on Biological Diversity (CBD) in 1992 and 1994, respectively. The direct economic benefits of biodiversity run into trillion of dollars per year (Constanza *et al.* 1997). Major threats to biodiversity arise from loss of habitat, deforestation, inappropriate water and agricultural management and natural disaster. Homestead is an ideal site for the preservation and participatory conservation of plant species. Homesteads can play as a centre for *in-situ* or *ex-situ* conservation of southwestern native species as well as introducing species from other parts of Bangladesh, which would add value in the homestead gene pools, organismal and ecosystem diversity. Homesteads obviously can promote *in-situ* conservation and gene bank of local and indigenous fruit and vegetable species of Bangladesh. Therefore, the study again listed a good number of species for conservation with various attributes.

5.19.1 Homestead plant species deterioration compared to the past

Different plant species were identified which were deteriorating or diminishing gradually from the homesteads compared to the past. The respondent of this study had given different opinion from their memories. It was found that the following major and dominant species were decreasing compared to the past. Some of major species are given in Table-44. This is noticeable for the farmers to ensure available fruit species in their homesteads. These species as a whole are listed including rank and relative rank and shown in Appendix -8.

Different other species were located but their abundance value was very insignificant. According to respondents' opinion, these species are decreasing from the homesteads compared to the past. The local and scientific names of these species are presented in Table-44.1.

Table-44. Homestead plant species deterioration compared to the past in the study areas of the southwestern coastal region of Bangladesh according to respondent opinion.

English name	Scientific name	Respondent opinion regarding species changing over time			
		Species decreasing 2008		Status decreasing 1988	
		Rank	RR*	Rank	RR*
Palmyra palm	*Borassus flabellifer*	87	36	120	50
Date palm	*Phoenix sylvestris*	80	33	131	55
Sundri	*Heritiera fomes*	75	31	136	57
Wood nut	*Diospyros blancoi*	71	29	119	49
Jamun	*Syzygium fruticosum*	68	28	116	48
Indian buch	*Pongamia pinnata*	62	26	109	45
Mango	*Mangifera indica*	60	25	120	50
Coconut	*Cocos nucifera*	53	22	101	42
Tamarind	*Tamarindus indica*	50	21	98	41
Custard apple	*Annona squamosa*	50	29	97	40
Country fig	*Ficus hispida*	40	21	50	21
Neem	*Azadirachta indica*	40	17	79	33
Water lily	*Nymphaea nouchali, N. pubescens*	48	20	96	40

* RR-Relative rank of the respondent opinion.

Table-44.1. List of minor non-woody plant species of the homesteads which are diminishing compared to the past in the study areas of the southwestern coastal region of Bangladesh.

Local name	Scientific name	Local name	Scientific name
Ban-kathal	*Artocarpus chama*	Hargoza	*Acanthus ilicifolius*
Bhat	*Clerodendrum viscosum*	Hoglapata	*Typha angustata*
Binnachopa	*Vetiveria zizanioides*	Lajjyabati	*Mimosa pudica*
Wild jute	*Corchorus aestuans*	Murta	*Schumanianthus dichotomus*
Bontulsi	*Ocimum basilicum*	Nalhagra	*Phragmites karka*
Buno chaia	*Aerva lanata*	Shati	*Curcuma zedoaria*
Dal ghas	*Hygroryza aristata*	Thankuni	*Centella asiatica*
Kalialata	*Derris trifoliata*	Tit begun	*Solanum nigrum*
Gol pata	*Nypa fruticans*	Telakucha	*Coccinia cordifolia*

5.19.2 Threatened and rare species need to be conserved

Some species which need to be conserved immediately in the study areas are shown in Table-45. The species may be conserved *in-situ* in the homesteads in this region. i) *Syzygium cuminii*, ii) *Casuarina equisetifolia*, iii) *Momordica cochinchinensis*, iv) *Sapium indicum*, v) *Calamus tenuis*, vi) *Psophocarpus tetragonolobus*, and vii) *Abroma augusta*.

Conservation, obviously is a vital issue for feeding the present and future generation for the survival of all living entity in the biosphere. In well managed homesteads, landscapes often tree-dominated showed promise for biodiversity conservation. Homestead vegetation which are maintained by the rural community should be one possible strategy for biodiversity conservation in a populous country like Bangladesh. Therefore, three main conservation activities were needed such as: i) awareness building; ii) protection of existing rare and uncommon species; and iii) fruit tree improvement through breeding, propagation and crop husbandry. Overlaying all of these activities are the inclusion of local communities in the process, who are the ones to retain these species in home gardens at the first place, and the stakeholders who will determine whether home gardens indeed act as long-term repositories for biodiversity conservation.

Table-45. Threatened and rare species need to be conserved in the study areas of the southwestern coastal zone of Bangladesh.

Local name	English name	Scientific name
Abeti	Cane	*Flagellaria indica*
Atafal	Custard apple	*Annona squamosa*
Bantula	---	*Hibicus moschatus*
Bhola	---	*Cordia dichotoma*
Buno Karol	Teasle gourd	*Momordica cochinchinensis*
Couphal	Cowa	*Garcinia cowa*
Chatian	Devils tree	*Alstonia scholaris*
Dakur	--	*Cerbera odollam*
Hijal	Indian oak	*Barringtonia acutangula*
Kamranga sheem	Winged bean	*Psophocarpus tetragonolobus*
Mamakala	---	*Sarcolobus carinatus*
Mewa kathal	---	*Annona muricata*
Mamakathal		*Morinda angustifolia*
Mouseem	Sword bean	*Canavalia gladiata*
Nagmani	---	*Wissadula periploci folia*
Pechigab	---	*Diospyros malabarica*
Royna	Rohina	*Aphanamixis polystachya*
Sada Dhutra	---	*Datura metel*
Urigab/Bangab	---	*Diospyros montana*
---	---	*Hyptis capitata*

5.20 Comprehensive strategies needed to forward homestead biodiversity

Increasing sustainable production in the homestead is an integral approach. It encompasses combination and integrated utilization of natural resources. An eco-friendly manner and approaches are mandatory to practise in the homestead biodiversity management. A comprehensive plan has been suggested in view of homestead biodiversity for sustainable production (Box-2). The main features of this plan logically framed incorporating experience and farmers opinion of the study are given as : i) The massive plantation (fast-growing and multipurpose timber-yielding and fruit-yielding species, quality seedlings, commercial vegetables cultivation, local and minor fruit, medicinal and wild species), ii) supply of agricultural inputs (organic composting, fertilizers, quality seeds, pesticides, power tiller), iii) modern cultivation (better seed preservation techniques, improved varieties, increased production), iv) financial assistance (ready cash, micro credit, bank facility), v) homestead space utilization (landscape planning, technical assistance, training), vi) homestead management (tree management and improvement), vii) poultry-livestock and fisheries management (technical support, training), viii) disaster management (flood control, disaster preparedness, coping capacity), ix) drinking water (tube well sinking, arsenic and salt-free water, preservation of rain water), x) minimization of salinity (maintenance of sluice gates, embankments and channels, canals re-excavation), xi) government policy (adoption and monitoring, input support, safety net).

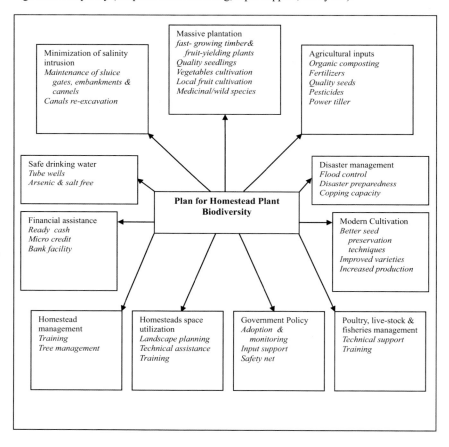

Box-2. Comprehensive steps needed to be taken for homestead biodiversity utilization and conservation in Bangladesh.

5.21. A typical model for sustainable homestead production

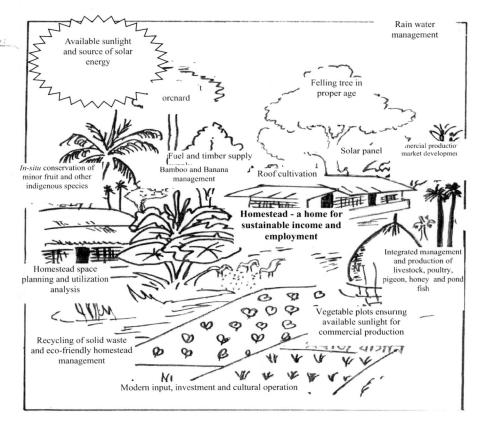

Box-3. A framework of homestead production through eco-friendly utilization and conservation of natural resources.

CHAPER 6: CONCLUSIONS AND RECOMMENDATIONS

CONCLUSIONS

Based on the findings and discussion the following conclusions are drawn:

6.1 Plant species identification and characterization

6.1.1) A total of 189 tree species were identified in the Southwestern coastal region. The existence of these large number and wider range of different plant species growing in the different ecological habitats in the homesteads proved the richness of species in terms of organismal and ecological diversity.

6.1.2) The present investigation found much more diversity and richness in vegetable-yielding species in the southwestern coastal homesteads. Out of the identified 62 vegetables, Chilli, Yam and Banana as low-cost vegetables grow well in the homesteads in the study areas.

6.1.3) The plants were systematically organized into 73 families of trees species and 18 families of vegetable-yielding species which include a wider range of diversity of dicot and monocot plants.

6.1.4) The number of tree and vegetables species varied as per salinity level of the study areas. A total of 29 species were characterized as moderate to strongly saline-tolerant and 19 as less saline-tolerant species. Salinity causes damage or death of some economically important plant species. Saline-tolerant cultivars and species are needed to be introduced for increasing overall homestead production.

6.1.5) On an average, a good number of tree species existed in the homesteads for sustainable production. The landless farmers accommodated a larger number of trees for their livelihood and survivability. Some of the dominant species in the homesteads of the Hindu and Rakahain communities were needed to be improved and popularized.

6.1.6) Some low-cost minor popular fruits and vegetables were grown which possessed economic value. These are: *Musa paradisica, Musa paradisica, Tamarindus indica, Sonneratia caseolaris, Spondias pinnata, Diospyros blancoi, Capsicum frutescens and Eryngium foetidum.*

6.2 Relative prevalence and diversity indices of tree species

6.2.1) The identification of the most prevalent and dominant timber-yielding species is a significant finding of the study. The species were: *Albizia richardiana, Swietenia macrophylla*, and *Samanea saman*. On the other hand, most preferable top ranking vegetables species were *Lagenaria sicerari, Carica papaya, Dolichos lablab*, and *Cucurbita maxima*.

6.2.2) Only a fewer number of timber-yielding and vegetables species were being preferred for plantation which seemed to be harmful for enhancing total biodiversity and food security of this southwestern region.

6.2.3) The alarming gap in terms of abundance between and among timber-yielding and fruit-yielding including other species may result a negative relationship in terms of organismal and ecosystem diversity in the long run which were influencing the livelihood of the inhabitants.

6.2.4) The species biodiversity indices of all farm categories were higher in the southern zone where plant biodiversity increases as the farm size increases. The diversity and abundance of timber-yielding species of all farm categories were higher than those of other species.

6.2.5) There is a positive correlation of diversity indices and equitability where diversity indices decreased with equitability along with the diminishing size of the farm category.

6.3 Changing pattern of homestead plant species

6.3.1) The trend of changing pattern of plant species, especially the proportion of timber-yielding and fruit-yielding trees in each homestead was found to be unbalanced. However, the farmers preferred to preserve and protect fruit-yielding trees for their long gestation period and harvesting fruits. Massive plantation of fewer number of selected timber-yielding species has been changing the past scenario which may create impediment on the overall development of homestead plant biodiversity.

6.3.2) Felling trend showed that landless farmers cut trees in short gestation period due to many socio-economic stresses and to meet the family emergency needs. Contrarily, large farmers kept old-aged trees and felled them in long gestation period which was not eco-friendly to increase homestead production and proper utilization of the micro-sites of the homesteads.

6.3.3) The positive correlation between felling trend and farm size was noted where average felling trend increased with increase of farm size. Increasing awareness among the farmers to fell trees in proper age limit would improve homestead productivity and proper space utilization as well as homestead biodiversity.

6.4 Homestead biodiversity regeneration

6.4.1) The landless and marginal farmers were found bias for planting fruit-yielding species because of household consumption, food security and sales for regular cash income generation to meet household daily expenses.

6.4.2) Rural homesteads of southwestern zone is still rich in indigenous and local agricultural cultivars. Hence the studied areas could be considered as ex-*situ* and *in-situ* gene banks which may play an immense role in agrobiodiversity regeneration. Therefore, combination of farmers choice along with species diversity shall have to be seriously considered for biodiversity conservation.

6.4.3) Different indigenous techniques were used for preservation of majority seeds. Farmers were not familiar with modern preservation and storage techniques. This needs to be promoted for conservation and maintenance of agrobiodiversity.

6.5 Major management practices and problems faced by the farmers

6.5.1) Farmers of the study areas usually use local and indigenous practices for management of their homestead production. Common management techniques and practices found in this region were organic manuring, chemical fertilizers, earthing-up, thinning, pruning and fencing.

6.5.2) The major problems identified in the management and production of homestead plant biodiversity were lack of advanced knowledge and technology, pests and diseases infestations, improper homestead space planning and utilization, lack of maintenance of embankment, salinity intrusion, low productivity, shortage of labor, recurring natural disasters, canals siltation, land scarcity, and social problems.

6.6 Homestead income, food security and contribution of timber and vegetable

6.6.1) Homestead production system was found to contribute **one third** of the total income of the household.

6.6.2) Trees significantly contributed diversely in the life supporting system especially food, fruits and vegetables and cash earning along with other usage. Some of the naturally growing plants considered as **"life supporting species"** provided enormous opportunity of food security to the marginal farmers, especially during the crisis period. The dependencies on diversified off-farm activities for the farmer were increasing in the rural areas which need to be promoted as an alternative source of income. Reduction of poverty through increasing homestead production and daily income needs to be paid attention for commercial vegetable and fruit production.

6.6.3) One third of the marginal farmers' **food security** was insecured which was a big threat to the farmers of this region.

6.6.4) A marginal family can ensure his major annual income from the following fruit-yielding species: Mango, Coconut, Palmyra palm, Date palm, Tamarind and Jujube. Besides, timber-yielding species Chambol, Mahogony and Raintree. These species are treated as the poor man's species for **"poverty reduction"**.

6.7 Strategies for homestead sustainable production and plant biodiversity conservation

6.7.1) Ensuring women's participation in decision-making and policy formulation is essential for homestead plant biodiversity conservation and regeneration and their rights and access to the plant resources need to be reserved.

6.7.2) The species which need to be conserved for *in-situ* and ex-*situ* environment are: i) *Syzygium fruticosum*, ii) *Casuarina equisetifolia*, iii) *Momordica cochinchinensis*, iv) *Sapium indicum*, v) *Calamus tenuis, vi) Psophocarpus tetragonolobus, and* vii) *Abroma augusta*.

6.7.3) Proper scientific management practices along with the use of surface water need to be introduced for sustainable homestead production.

6.7.4) The following comprehensive plan has been suggested in view of homestead biodiversity for sustainable income and production: i) massive plantation (fast-growing and multipurpose timber-yielding and fruit-yielding species, fruit-yielding species seedlings, commercial vegetables cultivation, local and minor fruit, medicinal and wild species), ii) supply of agricultural inputs, iii) modern cultivation, iv) financial assistance, v) homestead space utilization vi) homestead management, vii) disaster management, viii) minimization of salinity, and ix) government policy.

6.7.5) An eco-friendly approach is mandatory to practise in the homestead's biodiversity management. Long and short range strategic steps are logically framed according to experience and farmers' opinion gathered in the present investigation.

RECOMMENDATIONS

The study derives the following recommendations:

6.8.1) A comprehensive homestead space planning, proper scientific management of plants using diversified agricultural and forest cultivars are needed to promote homestead plant biodiversity and their productivity.

6.8.2) A policy guideline is to be developed for production and distribution of the quality planting materials, especially for timber-yielding, fruit-yielding, medicinal and vegetables.

6.8.3) Government should launch a program on **"a homestead and an ideal farm"** which would foster the utilization and conservation of homestead plant biodiversity.

6.8.4) Steps should be taken to develop village based plant inventory and preservation of specimens through engaging teachers and students of secondary schools and colleges of the respective localities.

6.8.5) Community **"seed bank"** is to be established from which farmers shall collect and preserve available seeds/cultivars through improved selection method along with the technical help of GOs and NGOs. This will function as an on-farm maintenance of agro-biodiversity. This would ensure change of behavior towards modern technology by the individual farmer.

6.8.6) Research program needs to be undertaken to improve 'Sorjan' method which is the most popular production technology which is conducive for biodiversity improvement in the southern zone of Bangladesh.

6.8.7) Some species which are threatened need to be conserved and preserved. These species are i) *Syzygium fruticosum*, ii) *Casuarina equisetifolia*, iii) *Momordica cochinchinensis*, iv) *Sapium indicum*, v) *Calamus tenuis*, vi) *Psophocarpus tetragonolobus*, vii) *Abroma augusta*, viii) *Canavalia gladiata*, ix) *Diospyros montana*, x) *Morinda angustifolia* and *Hyptis capitata*.

6.8.8) Steps should be taken to support farmers through giving credit with minimum or no interest to foster homestead production system.

6.8.9) Long and short range strategic steps with logical frameworks for management and improvement of homestead production system while maintaining biodiversity and ecology are needed.

CHAPTER 7: REFERENCES

Ahmed, F.U. 1997. Minor fruits in homestead agro forestry. In: Agroforestry: Bangladesh perspectives. Alam *et al*. (eds.), APAN/NAWG/BRAC.PP. 165-169.

Ahmed, Z.U., Hasan, M.A., Begum, Z.N.T., Khondker, M., Kabir,S.M.H.,Ahmed,M. and Ahmed, A.T.A (eds.). 2009. Encyclopedia of Flora and Fauna of Bangladesh, Vol.1. Asiatic Society of Bangladesh, Dhaka.

Ahmed, M.F.U. 1999. Homestead agroforestry in Bangladesh: A case study of Gazipur District. MS Thesis, Dept. of Agroforestry & Environment, BSMRAU. Salna, Gazipur, Bangladesh.

Ahmed, Z. U. 2003. Banglapedia: National Encyclopedia of Bangladesh. Asiatic Society of Bangladesh.

Akhter, M. S., Abedin, M.Z., and M. A. Quddus. 1989. Why farmers grow trees in Agricultural fields: some thought, some results. In: Research Report 1988-89. On-Farm Research Division, Jessore. Bangladesh Agricultural Research Institute. PP. 161-178.

Alam, M. S. and K. M. Masum. 2005. Status of homestead biodiversity in the offshore island of Bangladesh. Research Journal Agriculture and Biological Science 1(3). PP. 246-253.

Alam, M. K., M. Mohiuddin M. and S. R. Basak. 1996. Village trees of Bangladesh: diversity and economic aspects. Bangladesh Forest Research Institute, Chittagong, Bangladesh.

Alam, Q.M., M.M., Ullah, M.M. and M.A. Rashid. 1995. Economic condition of small farm household in selected districts of Bangladesh. Bangladesh Journal Agriculture Research, Joydevpur, Gazipur, Bangladesh.

Ara, H., B. Khan and S. N. Uddin. 2013. Red Data Book of Vascular Plants of Bangladesh, Volume-2. Bangladesh National Herbarium, Ministry of Environment and Forests, Chiriakhana Road, Mirpur, Dhaka, Bangladesh.

Anam, K., 1999. Homesteads agroforestry in the level Barind tract: A diagnostic study. MS Thesis, Dept. of Agro. and Env., BSMRAU. Salna, Gazipur, Bangladesh.

Arefin, M. K., M. M. Rahman, M. Z. Uddin and M.A. Hasan. 2011. Angiosperm Flora of Satchari National Park , Habiganj, Bangladesh. Bangladesh Journal of Plant Taxonomists. 18(2):117-170. 2011.

Atikullah , S.M. and M. Hassanullah. 1997. Factors Affecting Role of Extension Workers of Government and Non-government Organization. Bangladesh

Journal Training and Development 13. (122), 200:149-158, Graduate Training Institute. Mymensingh, Bangladesh.

Bangladesh Meteorological Department. 1983-2007. Yearly rainfall and temperature of Patuakhali and Khepupara Station. Meteorological Department, Climate Division, Govt. of Bangladesh, Begum Rokeya Sarani, Agargaon, Dhaka, Bangladesh.

BBS. 2011. Statistical pocket book of Bangladesh. Bangladesh Bureau of Statistics, Ministry of Planning, Bangladesh. PP. 37.

Basak, N. R. 2002. Study of composition of trees in homesteads at different ecological zones in Bangladesh. MS Thesis, Dept. Agri. and Env., BSMRAU, Salna, Gazipur, Bangladesh.

Basher, A. 1999. Homegarden agroforestry: impact on biodiversity conservation and household food security, a case study of Gazipur district. MS thesis, CIEDS, Noragic Agri. Univ., Norway.

BBS. 2001. Population Census, National Report, Planning Division, Ministry of Planning, Govt. of Bangladesh, PP. 93-97.

BBS. 2001. Statistical yearbook of Bangladesh. Bangladesh Bureau of Statistics, Ministry of Planning, Bangladesh.

BBS. 2002. Statistical year book of Bangladesh. Bangladesh Bureau of Statistics, Ministry of Planning, Govt. of Bangladesh.

BBS. 2006. Statistical pocket book of Bangladesh. Bangladesh Bureau of Statistics, Ministry of Planning, Bangladesh. PP. 37.

BBS. 2007. Statistical pocket book of Bangladesh, Bangladesh Bureau of Statistics, Ministry of Planning, Govt. of Bangladesh.

BBS. 2007. Report of the household expenditure survey 2005. Bangladesh Bureau of Statistics, Ministry of Planning, Govt. of Bangladesh, Dhaka, Bangladesh.

Constanza, R., d'Arge, R., de Groot, R.,Fabes, S.,Grano,M.,Hannon,B.,Limburg,K.,Naeem,S.,O'Neil,R.V.,Pareulo,J.,Raskin,R.,Sulton,P. and van der Belt, M. 1997. The value of world's ecosystem services and natural capital.

DFID, 2002. Department for International Development. Biodiversity a crucial issue for the world's poorest. ISBN 186-192-341-4.

Doglas, J. S. and R. A. De J. Hart.1973. Forest Farming. Watking, London, U.K.

FRMP. 2000. Forestry Resources Management Project. Technical Assistance Component. Integrated Forest Management Plan for the Patuakhali Coastal Aforestation Division, Forest Department, Dhaka.

Heywood, V. H . and R. T. Watson. 1995. Global Biodiversity Assessment. Published for the United Nations Environment Programme. Cambridge University Press. PP. 5-105.

ICRAF.2010. Path to Prosperity through Agroforestry, ICRAFs Corporate Strategy 2001-2010. International Centre for Research in Agroforestry.

Islam, Sk.A., M. A. Quddus. Miah and M. A. Habib. 2013. Diversity of Fruit and Timber Trees in the Coastal Homesteads of Southern Bangladesh. J. Asiat. Soc. Bangladesh. Sci. 39(1): 83-94, 2013.

IUCN. 1980. International Union for Conservation of Nature and Natural Resources. World Conservation Strategy. IUCN, Gland, Switzerland.

Kabir , M. E. and E. L. Webb. 2007. Can Homegardens Conserve Biodiversity in Bangladesh. Biotropica. 40 (1): 95-103. Natural Resources Management, School of Environment, Resources & Development. The Asian Institute of Technology. Klong Luang, Pathum Thani, Thailand.

Kamal, A. M. and A. B. M., Rahman. 2005. District Information: Patuakhali and Barguna. Program Development Office, Integrated Coastal Zone Management. Water Resource Planning Organization, Ministry of Water Resources, Government of Bangladesh. PP. 5.

Kumar, B.M., S.J.George and S. Chinnamani. 1994. Diversity, structures and standing stock of wood in the homesteads of Kerala in peninsular India. All India Coordinated Research on Agroforestry, College of Forestry, Kerala Agri. Uni., Vellanikkara, India, Agroforestry systems. 25(3): 243-262.

Kumar, U. and J. A. Mahendra. 2000. Biodiversity Principles and Conservation. Department of Botany, Govt. Dungar College, Bikaner 334 001, India. PP. 53-68.

Latif, M. A., Alam M. K. and M. Millat-e-Mustafa. 2001. Floristic diversity, growth statistics and indigenous management technique of traditional home gardens in Bangladesh. Final report of a contract research project of BRAC, BFRI and IFESCU, Bangladesh.

Mandal, A.S. 2008. World food security: the challenges of climate change and bio-energy: Bangladesh perspective. Food for all Saranika (Souvenir) 16 October, 2008 in the observation of World Food Day, BAU, Mymensingh.

Mandal, A. S. 2003. Trend in rural economy in Bangladesh: Issue and strategy for development. In: Natural Resource Management: Towards better integration. A. R. Rahman, N. Haque, and D. Mallick (eds). Bangladesh Centre for Advanced Studies and Department for International Development, Bangladesh.

Mannan, A. S. 2000. Plant biodiversity in the homesteads of Bangladesh and its utilization in crop improvement. Ph.D. Thesis, Genetics and Plant Breeding Dissertation, BSMRAU, Salna, Gazipur, Bangladesh. PP.36.

McNeely, J. A., K.R., Miller K.R., Reid. W., Mittermeier R. and T, Wrner.1990. Conceiving the World Biodiversity. Gland, Switzerland and Washington D.C., World Conservation Union and World Resource Institute.

Miah, M.G., Abedin, M.Z., Khair, A.B., M.A., Shahidullah M.A. and A.J.M.A. Baki. 1990. Homestead plantation and household fuel situation in the Ganges Flood Plain in Bangladesh. In: Abedin et al. (eds.), Homestead plantation and agroforestry in Bangladesh. Proceed. National Workshop held on 17-19 July, 1988 at Joydevpur, Gazipur, Bangladesh.PP.106-119.

Miah, G. M. and M. M. Ahmed. 2003. Traditional Agroforestry in Bangladesh: Livelihood activities of the rural households. A poster presented at the XII World Forestry Congress, Canada, held in September, 2003.

Miah, G. M. and N, Bari. 2001. Traditional Agroforestry in Bangladesh: Livelihood activities of the rural households. A poster presented at the XII world forestry congress, 2003.

Millat-e-Mustafa, M. and A.K. Osman Haruni. 2002. Vegetation characteristics of Bangladesh homegardens. Forestry Project, Intercooperation, Rajshahi, Bangladesh and Institute of Forestry & Environmental Sciences. University of Chittagong, Bangladesh.

MoEF. 2008. Ministry of Forest and Environment. Bangladesh Climate Change Strategy and Action Plan, 2008. Ministry of Environment and Forestry, Dhaka, Bangladesh.

Moran, S.A. 1984. Systematic and regional biography. Van Nostrand Reinhold Company, New York, USA. PP 334.

Nishat, A., Imamul Huq, S.M., Shuvashish, P.B., A.H.M., Ali Reza. and A.S.Moniruzzaman Khan. A.S. 2002. Bio-ecological Zones of Bangladesh, IUCN Bangladesh, Dhaka, Bangladesh.

Novib. 1996. Bangladesh country policy document: Summary proceedings of the mid-term review workshop. Dhaka, Bangladesh.

Rahim, M. A. and M. A. Haider. 2000. Multistoried cropping for sustainable land use, vertical yield enhancement, biodiversity, agricultural conservation in Agroforestry system. BAU, Mymensingh and Training Officer, PKSF, Dhaka.

Rashid, M.A., Rahman, M. Hussain T. S., and M. M. Rahaman. 2004. Indigenous vegetables in Bangladesh. International Conference on Indigenous Vegetables and Legumes. Prospectus for Fighting Poverty, Hunger and Malnutrition, International Society for Horticultural Science.

Rahman, A. K. A. 1992. Wetlands of Bangladesh. Bangladesh Centre for Advanced Studies & Natural Conservation Movement, Dhaka, Bangladesh.

Rahim, M. A. and M. A. Haider. 2000. Multistoried cropping for sustainable land use, vertical yield enhancement, biodiversity, agricultural conservation in Agroforestry system. BAU, Mymensingh and Training Officer, PKSF, Dhaka.

Razzak, M. A., M. I. A. Howlader, S,. Rafiquzzaman and S. C. Dam. 2001. Pattern Basis Vegetable Production in Homestead Area (Lebukhali Model). On-Farm Research Division, BARI, Patuakhali. p. 5.

Saha, S. 2001. Homestead space planning: a field guide. Strengthening Household Access to Bari Gardening, CARE-Dinajpur, Bangladesh.

Sheikh, K., Tahira-Ahmed and M.A., Khan. 2002. Use, exploration and prospects for conservation: people and plant biodiversity of Naltar valley, northwestern Karakorums. Dept. of Bio. Sci., Wildlife and Biodiversity Research Group. Biodiversity and Conservation 11(4): 715-742.

SRDI. 2000. Soil Resource Development Institute. Soil Salinity in Bangladesh. Govt. of the People's Republic of Bangladesh, Dhaka, Bangladesh.

Sultana, A. 2004. Plant biodiversity in the homestead microsites in Ganges floodplain bioecological zone of Bangladesh. MS Thesis, Dept. of Agrf. & Env., BSMRAU, Salna, Gazipur, Bangladesh.

Uddin, M. S., Rahman, M. Z., M.A. Manan, S.A. Begum, A.F.M.F. Rahman and M.R. Uddin. 2001. Plant Biodiversity in the Homesteads of Saline Area of Southern Bangladesh. Plant Breeding Division and Seed Technology Division, BARI. Pakistan J. Biol. Sci. 5 (6): 71-714.

UNCED.1992. United Nations Conference on Environment and Development. Rio de Janerio, Brazil.

Wakley, E. and J. H. Momsen. 2007. Women and seed management: A study of two villages in Bangladesh. **Department of Human and Community Development, University of California, Davis, USA.**

Wezel, A. and S. Bender. 2002. Plant species diversity of homegardens of Cuba and its significance for household food supply. Agrof. System. International Nature Conservation, Institute for Zoology, Univ. of Greifswald, Germany. 57:9-49.

WRI/IUCN/UNEP. 1992. Global Biodiversity Strategy. WRI/IUCN/UNEP, Washington DC.U.S.A.

Vertical view of a homestead   A house of homestead   Woman in the homestead

Vegetable species growing on the   Paddy threshing in courtyard   Homestead approach road
homestead roof

Ornamental plant (*Bougainvillea*   Homestead - a shelter of   Pegeon cage in a homestead
spectabilis) in the homestead   poultry bird and goat

Homestead with houses, vegetables   Bottle gourd in the court-yard Pepey (*Carica papaya*) in a
and plants   (*Legenaria siceraria*)   homestead

Plate-1.Rural homesteads: utility and uses.

Palmyra palm leaves for roof/wall preparation

Palmyra palm lumber ready for sale

Hand fan made-up of palm leaves

Ripe Tal (*Borasus flabellifer*)

Green Tal as rural food and nutrition

Hanging fruits of Tetul (*Tamarindus indica*)

Aam (*Mangifera indica*) hanging from the roof top of a homestead

Green Coconut (*Cocos nucifera*) for sale

Ripe Coconut shells for sale

Hanging fruits of Kathal (*Artocarpus heterophyllus*)

Fruits of Jambura (*Citrus maxima*) on the roof top

Fruits of Kat badam *(Terminalia catappa)*

Plate- 2. Different uses of economic and profitable fruit-yielding species.

A Date palm tree (*Phoenix sylvestris*) in fruit bearing stage

Standing tree of Date palm (*Phoenix sylvestris*)

Date for sale in the market

Fruits of Betel-nut (*Areca catechu*)

Betel-nuts ready for sale in a country boat

Fermented and peeled Betel nuts

Bamboo (*Bambusa tulda*)

Poultry cage and winnows

Dala/Busket

Fruits of Amra (*Spondias pinnata*)

Fruits of Bilatigab (*Diospyros blancoi*)

False fruits of Chalta (*Dillenia indica*)

Plate-3. Different economic and profitable species and their uses.

A bunch of Atia kala (*Musa sylvestris*)

Alternative fishing boat made-up of Banana pseudostems

Banana pseudo stems and inflorescence for household consumption

Mature fruits of (*Musa paradisiaca)* called Anaji kala

Bunches of Kathali kala (*Musa paradisica*)

A plant of Polau pata (*Pandanus odorus*)

Fruits of Pechi gab (*Diospyros embryopteris*)

Hot Chilli (*Capsicum frutescens*) *cv.* 'Bombai'

Bilatidhaina pata (*Eryngium foetidum*)

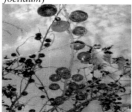

Fruits of Buno amra (*Spondias pinnata*)

Tiger fern (*Acrostichum aureum*), a salinity indicator

Fruits of Urmai (*Sapium indicum*)

Plate-4. Different economic and life supporting species.

Fruit and flower of
Stereospermum sp.

Fruits of Dakur (*Cerbera odollam*)

Fruits of Mewa kathal
(*Annona muricata*)

Fruits of Sada dhutra (*Datura metel*)

Fruits of Bantula (*Hibischus moschatus*)

Fruits of Mewa kathal
(*Annona muricata*)

Mamakala flower and fruits
(*Sarcolobus carinatus*)

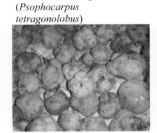

Fruits of Kamranga sheem
(*Psophocarpus tetragonolobus*)

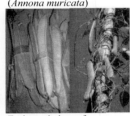

Fruits and plant of
(*Canavalia gladiata*) called
Mou sheem

A fruit of *Aegle marmelos*, known
as Bel

Fruits of *Artocarpus lacucha*,
known as Deuwa

Fruits of *Garcinia cowa*,
known as Cau

Plate-5. Different rare and uncommon species.

Two coloured flowers of (*Thevetia peruviana*) known Kalkephul

Nayantara (*Catharanthus roseus*)

Tulsi (*Ocimum sanctum*)

Jhumko Jaba (*Hibiscus schizopetalus*); and Common Jaba H. rosa-sinensis

Hasnahena (*Cestrum nocturnum*) and *Hyptis capitata*

Nagmoni (*Wissadula periplocifolia*)

Bamunhatti (*Clerodendrum indicum*) in bloom

A children holding a bunch of Kadam (*Neolamarckia cadamba*) flowers for sale

Apang (*Achyranthes aspera*)

Kewakata (*Pondanus foetidus*) harvested for mat preparation

Hargoza (*Acanthus ilicifolius*) a saline tolerant species

Hoglapata (*Typha elephantina*)

Plate-6. Diversified homestead species of Southwestern region of Bangladesh.

120

Mango seedlings (bare rooted)

Indigenous seedling raising

Vegetable seedling sale

Indigenous fruit maturing method

Indigenous seed preservation in earthen "*Motki*"

Prepared 'Sorjan' bed adjacent to a homestead

Guava management system

Top-dying of Date palm and Pummelo

Mango disease (Canker)

A bare homestead post cyclone Sidr

An uprooted tree of *Samanea saman* after cyclone Sidr 2007

A damaged tree of Betel nut (*Areca catechu*)

Plate-7. Seedling and seed raising techniques, fruit diseases and Sidr damage of homestead resources.

Haicha (*Alternanthera philoxeroides*), Helencha (E*nhydra fluctuans*) are being collected by the rural women and adolescent as a source of naturally grown vegetables.

Chaila (*Sonneratia caseolaris*)

Fruits of Jagga dumur (*Ficus glomerata*)

Fruits of Deshi dumur (*Ficus hispida*)

Floating Water lily in a pond (*Nymphaea pubescens*)

Peeled fruits of water lily for rural children

Water lily ready for sale in the market by the farmer

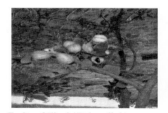  

Fruits of Jilapi (*Pithecellobium dulce*)

Fruits of Tit begoon (*Solanum nigrum*)

Mateyalu (*Dioscorea alata*)

Plate-8. Different 'life supporting species' as alternative food and income.

A rural farmer selling Jute leaves (*Corchorus capsularis*) in urban market

A village doctor displaying herbal medicine in his shop of a rural market

A rural farmer selling vegetables in urban market

Kachur lati: for sale in the urban market as an alternative food and income

Women winnowing paddy in the homestead

Date palm leaf mat preparation by a rural woman

Base of a felled Palm tree: log is sold as a source of income of the household

Indigenous earthen oven and fuel wood for homestead's energy

Indigenous device to drive away the birds for protecting Jambura

An eighty year old Rakhain women

Her son and daughter in-law

Rakhain house and indigenous hand loom

Plate-9. Important photographs on income and employment and a Rakhain family.

123

Focus Group Discussion with the
stakeholders

Discussion session with school
teachers

Supervisor in the seminar
session

Group discussion with villagers

School teachers in a photo
session

Teachers in seminar session

Co-supervisor in the FGD session
with local elites

Researcher in the Choulapara
site of the study

Researcher interviewing
with a farmer

A Rakhain elderly priest of
Nayapara village

A Muslim elderly priest
(*Imam*) in Kalisuri village

An elderly person of
Kalisuri village

Plate-10. Important photographs on FGD, seminar, field survey and elderly persons.

Appendix-1: Transect of the studied villages (Nayapara, Choulapara and Kalisuri)

Criteria	Nayapara	Choulapara	Kalisuri
1. Soil	Sandy, clay	Sandy-near river, clay	Sandy loam, clay
2. Land : High Medium high Low	30% 40% 30%	30% 50% 30%	40% 35% 25%
3. Drinking water	Tube wells 100 % Pond after SIDR, - 5 tube wells 3 before and 2 after Sidr	Tube wells 100 % Pond after SIDR, - 8 tube wells 3 before& 5 after Sidr	Tube wells 100 % 9 tube wells 8 before& 1 after Sidr
4.Saline water	Very high in dry season	High in dry season	No saline water
5. Homestead and house	High base (*uchu bhiti*) , tin-shade, *kacha*, one or two storied (ek-tala or do-tala)	High base (*uchu bhiti*) , tin-shade, kacha, one or two storied (ek-tala or do-tala)	High base (*uchu bhiti*) , tin-shade, kacha, one or two storied (ek-tala or do-tala)
5.1. Homestead High, Medium high & low	10%, 40 %, 50%	5%, 75 %, 20%	30%, 50 %, 20%
6. Court-yard	High and plain	Big and plain	Big and random
7.Village direction	East-West	North - South	North-South
7.1.Village periphery/ boundary	North – Alipur South-Kachapkhali East- Thanjupara West- Khajura canal	Like a island- Kachupatra, andharmanik & bagirkhal North - Chakamoiya South-Angarpara East- Jharakhali West- Chakamoiya	Village east bank is Aloki river which is connected with Tetulia river North- Bhatsala South-Kabirkati East- Aloki river West- Ponahura
8. Pond	Mini pond in each homestead, each pond overflows with saline water due to SIDR	Small, medium pond in each homestead, each pond overflows with saline water due to SIDR	Small, medium, big in each homestead.
9.Communication	Tricycle, Honda, Bus and walking	Boat, Engine boat, walking	Boat, Engine boat, Bus, Honda, walking
10.Literacy	Signed 80%	Signed 60%	Signed 85%
11. Primary School	1 registered	1 Government	2 Government
12.High school	--	--	1
13.Village market	--	--	1
14.Village tea stall shade	Tulatoli roadside	BWDB embankment	--
15.Village hospital	1 newly establishing	--	1 rural Dispensary
16.Water (rainy and dry season)	Sweet water (Misti pani)- Ashar-Kartik, in rainy season and apart from the season Highly saline (kachu gach -O-maray jai)	Sweet water (Misti pani)- Ashar-Kartik, in rainy season and apart from the season Highly saline	Sweet water (Misti pani)- Ashar-Kartik, in rainy and, no saline water
17.Vegetables	Lalsak, Lau, Shim, Barbati, Karola	Lalsak, Lau, Shim, Barbati, Karola	Lalsak, Lau, Shim, Barbati, Karala
18.Fruits	Aam, Baroi, Tetul, Kala , Papaya, Tal, Khejur	Aam, Baroi, Tetul, Kala , Pepey, Tal, Khejur	Aam, Tetul, Kala, Pepey,Gab, Tal, Khejur
19.Trees	Raintree, Mahogony, Chambol, Babla	Raintree, Mahogony, Chambol, Babla	Raintree, Mahogony, Chambol
20.Khana	155	209	305

Appendix -2: Questionnaire of the study

Homestead plant identification, utilization and conservation survey:

Questionnaire

Department of Botany

Jahangirnagar University, Savar, Dhaka

Sample code number:

Farmers category (√): i. Landless ii. Small iii. Medium and iv. Large

Family information

Household demographic information

Name   :   Village:
Upazilla :   District   :
Age   :   Education (class) :
Marital Status (√): i. Single ii. Married, iii. Widowed
Family member's: .... , M. ...., F................/Adult........., non-adult................

Farm and off-farm occupation and contribution to the livelihood

Farm occupation	Contribution (%)	Off-farm Occupation	Contribution (%)

Homesteads structure and land utilization

Homestead prepared in different faces (√): East/West/South/North.

Reason for choosing different faces while constructing of this homestead.

Spatial arrangement of plants species surrounding the homestead by visual observation (√).
i. Over utilized, ii. Full utilized, iii. Partially utilized iv. Under-utilized.

Total farm size of the household acre.., and distribution of land.
i. Homestead land ....acres, ii. Cultivated land ....….. acres.

Ownership pattern of land

Ownership pattern and sources of land	Tick mark	Amount (acres)
Paternal		
Purchase		
Rented		
Government		
Others		
Total=		

Causes of increasing/decreasing of farm land.
Present pattern of homestead land utilization.

Space/area Distribution	Area		Reason (increasing/ decreasing
	Local	Hectare	
Houses (shade)			
Court-yard			
Garden (vegetable, orchard, trees)			
Pond/dishes/wetlands			
Approach road			
Surroundings/others			

Identification and characteristics of homesteads plants species

Homestead tree species

Name of trees	Number					Total	Uses
	<5 years	6-10 years	11-15 years	16-20 years	>20 years		
Timber-yielding species							
Fruit-yielding species							
Medicinal and spice-yielding species							
Ornamentals species							
Naturally-growing							
Non-woody plants							

Cultivated, naturally growing vegetables

Name of vegetables	Summer	Winter	Year round	Uses/Characters

Contribution of homestead cultivated vegetables...........%, naturally-growing vegetables ....%, & purchase vegetable from rural markets ....% (100%)

Seed collection and preservation techniques

Please mention do you preserve seeds in the homesteads (√): Yes/No.

Major agricultural seeds collected and stored (vegetables, pulse, oil & others) in the homesteads.

Name of seeds	Variety/Species	Stored/preservation techniques	Effectiveness/ demerits

Trend of homestead plantation during the last 10 years

Plant species	Nos.	Age	Purpose/likings
Fruits-yielding			
Timber-yielding			
Others			

Source of planting materials/seedlings

Sources	Name planting materials/number	Unit price (Tk)	Remarks
Government nurseries			
NGO nurseries			
Self/own production			
Rural markets			
Relatives			
Others			

How many trees that have been fell during the last 10 years?

Name of Trees	Years	Age	No	Purpose/uses
Timber-yielding				
Fruit-yielding				
Others				

Do you think that trees have been affecting your livelihood support and any impact in income generation of the family and how?

Changes	How
Livelihood support	
Ecosystem/Environment	

List of major trees and other plants that are not available in the homesteads at present but was available past.

Fruit/timber/fuel-yielding and medicinal Plants

Name of plants	Name of plants	Threatened/ Uncommon	Endangered/ Rare	Remarks/ Causes
Timber /Fuel-yielding				
Fruit Yielding				
Medicinal				
Non-timber plants				
Others				

Homestead income from different sources

Sources of income	Unit price	Price (Taka)	%	Comments
Homesteads income				
Field crops income				
Service income				
Business				
Livestock				
Labor sale				
Handicrafts				
Others				
Total				

Homestead expenditure incurred by different categories

Sources of income	Unit price	Price (Taka)	%	Comments
Educational expenses (monthly/yearly)				
Daily different expenses				
Rice purchase/self				
Fuel (oil, wood)				
Edible oil				
Clothes				
Medical treatment				
Religious festivals				
Others				
Total				

Food security and relative contribution

Do you consider homesteads as a source of income: (√) Yes/no.
Why/Causes 1:

What is the status of household food security (√)?

10 years before				Present time			
Good	Stable	Poor	Fragile	Good	Stable	Poor	Fragile

How many meals do you have per day normally (√): 1…, 2…, 3.

Did you store food grains for future and food security in future? (√) Yes/No.

If yes please mention name of stored food grains and time of store.

Stored food items	Amount (Kg)	Time of storage	Time of consumption	Why store

Please mention the name of economic and multipurpose tree species

Name of species	Rank	Uses

Cultural and management practices taken for homestead production

Management practices	Tree species	Amount	When	How	Comments
Use of Organic fertilizers					
Use of Chemical fertilizers					
Use of pesticides					
Earthling-up					
Thinning					
Pruning					
Fencing					
Others					

Problems faced in homesteads plantation, and management.

Problems	Solutions

Did you receive any training on agriculture/forestry or horticulture?

Name of training	Year	Organization	Skill learned /Benefits

Role of women in decision making regarding plantation, vegetable cultivation, expensed in homesteads ....%. any opinion regarding gender role.

Please mention necessary steps need to be taken for increasing production and economic enhancement of the homesteads.

Any suggestions for the improvement of homestead farm and production.

Any other information, which would be useful in conserving homestead plants diversity.

Signature   Date

Appendix-3: List of homestead plant species identified and characterized in the study areas of the southwestern coastal region of Bangladesh.

Family	al name	English name	Scientific name	Uses
1. Acanthaceae	Kalomegh[6]	Creat	1) *Andrographis paniculata*	Hm,Mv
	Hargoza[6]	Holy-leafed acanthus	2) *Acanthus ilicifolius*	Hm,Fw,F
	Basak[6]	Malabar nut	3) *Justicia adhatoda*	Hm, Mv
2. Amaranthaceae	Buno chai[6]	-----	4) *Aerva lanata*	Hm
	Helencha[6]	Alligator weed	5) *Alternanthera philoxeroides*	F, Mv
	Apang[6]	Prickly Chaff flower	6) *Achyranthes aspera*	Hm, F
3. Amaryllidaceae	Rajanigandha[6]	Tube rose	7) *Polianthes tuberosa*	Fr
4. Anacardiaceae	Amra[1]	Hogplum	8) *Spondias pinnata*	Fr,Mv, J, Fw
	Kapila/Giga [2]	Wodier	9) *Lannea coromandelica*	Hp,P, F, Mv ,Fw
	Buno amra[5]	Wild hogpalm	10) *Spondias dulcis*	A & BF, Fr,T
	Aam[1]	Mango	11) *Mangifera indica*	Fr,Fw,Fe, Mv, J
5. Angiopteridaceae	Dhekirsak[6]	Edible fern	12) *Diplazium esculentum*	Vg, Mv
6. Annonaceae	Sarifa[1]	Custard apple	13) *Annona squamosa*	Fr, Mv, Fw
	Debdaru[4]	Most tree	14) *Polyalthia longifolia*	Fw,P, Fw
	Atafol[1]	Bullocks heart	15) *Annona reticulata*	Fr,Mv, Fw
7. Apiaceae	Thankuni[6]	Pennywort	16) *Centella asiatica*	Mv, Vg
8. Apocynaceae	Nayantara[4]	Periwinkle	17) *Catharanthus roseus*	Fr,Rf
	Karali[4]	Lucky nut	18) *Thevetia peruviana*	Fr, Rf, P, F
	Karabi[4]	Oleander	19) *Neium indicum*	Fr, Rf
	Chatian[2]	Devil's tree	20) *Alstonia scholaris*	At,Fw,H,T
9. Araceae	Bis/Kanta kachu[6]	-----	21) *Lasia spinosa*	Hm
10. Arecaceae	Tal [1]	Palmyra palm	22) *Borassus flabellifer*	Fr, J, Hf, M,T,P
	Khejur[1]	Date palm	23) *Phoenix sylvestris*	Fr,J,Fw,Iy, F,P
	Hetal[1]	-----	24) *Phoenix paludosa*	A & BF,Fw,P
	Narikel[1]	Coconut	25) *Cocos nucifera*	Fr, D, Fw, F, H
	Gol pata[6]	-----	26) *Nypa fruticans*	F,H, F
	Bet[6]	Cane	27) *Calamus tenius*	Ir,H,Fr
	Supari[1]	Betel nut	28) *Areca catechu*	Fr, Fw, F, P
11. Asclepiadaceae	Mamakala[6]	----	29) *Sarcolobus carinatus*	Fr, Mv
	Akond[3]	Swallow wort	30) *Calotropis gigantea*	Hm,Fw
12. Asteraceae	Bannalata[6]	Hempvine	31) *Mikania cordata*	Hm, F
	Ganda/Gada[6]	Marigold	32) *Tagetes patula*	Fr, Rf
13. Averrhoaceae	Kamranga[1]	Gooseberry	33) *Averrhoa carambola*	Fr, Mv, Fw
14. Bignoniaceae	Gorshingiah[5]	Mangrove trumper tree	34) *Dolichandrone spathacea*	Fw,P, T
	Nauasonail[2]	-----	35) *Oroxylum indicum*	Fw,P, T
15. Bombacaceae	Simul tula[2]	Silk cotton	36) *Bombax ceiba*	Iy, Mv, T
	Kat tula[2]	Kapok-tree	37) *Ceiba pentandra*	Iy, H, T
16. Boraginaceae	Bahal[5]	-----	38) *Cordia dichotoma*	T,Fw,Hv
	Hatisur[6]	Heliotrope	39) *Heliotropium indicum*	Mv, F
17. Brassicaceae	Bansarisha[6]	-----	40) *Rorippa indica*	Hm ,F
18. Bromeliaceae	Anaras[6]	Pine apple	41) *Ananas comosus*	Fr,Mv
19. Caesalpiniaceae	Gulmohur [4]	Flame tree	42) *Delonix regia*	Fr, Fw,Mv,T
	Sonail[2]	Indian laburnum	43) *Cassia fistula*	Fr,Fw, P, T
	Dadmordon[6]	Ringworm bush	44) *Senna alata*	Mv
	Tetul[1]	Tamarind	45) *Tamarindus indica*	Fr,Mv, T, Fw
20. Cannaceae	Kalabati[6]	China lily	46) *Canna indica*	Fr, Rf
21. Casuarinaceae	Jhau[4]	Seef wood	47) *Casuarina littorea*	P,H,Fw
22. Chenopodiaceae	Bathua[6]	Lambs qarter	48) *Chenopodium album*	Hm,Vs
23. Clusiaceae	Gaitta[5]	Borneo mahogany	49) *Calophyllum inophyllum*	Fw,At, P,F
	Couphal[1]	Cowea	50) *Garcinia cowa*	Fr, F, Mv
24. Combretaceae	Arjun[3]	Malabar nut	51) *Terminalia arjuna*	Mv,T, B, At
	Bohera[3]	Belleric myrobalan	52) *Terminalia bellerica*	Hm, Mv, Fw

131

	Hartaki[3]	Black myrobalan	53) *Terminalia chebula*	Hm,Mv
	Katbadam[2]	Indian almond	54) *Terminalia catappa*	Fw, Fr, T, Iy,H
25. Convolvulaceae	Kalmi lata[6]	Swamp cabbage	55) *Ipomoea aquatica*	Vg, F,F
	Sarna lata[6]	Giant dodder	56) *Cuscuta reflexa*	Mv, Hm
	Dholkalmi[6]	-----	57) *Ipomoea fistulosa*	F, F
26. Crassulaceae	Pathar kuchi[6]	Life plant/Floppers	58) *Kalanchoe pinnata*	Mv,
27. Cruciferae	Banmula[6]	Wild radish	59) *Raphanus raphanistrum*	Hm
28. Cucurbitaceae	Telakucha[6]	Ivy gourd	60) *Coccinea cordifolia*	Mv, Hm
29. Cyperaceae	Helipata[6]	-----	61) *Cyperus tagetiformis*	F, H, M
	Chachkata[6]	-----	62) *Eleocharis acutangula*	F
30. Dilleniaceae	Chalta[1]	Indian dillenia	63) *Dillenia indica*	Fr,Mv,A, T
31. Dioscoreaceae	Mateyalu[6]	Winged yam	64) *Dioscorea alata*	Vg
	Gach alu[6]	-----	65) *Dioscorea deltoidea*	Fr, Mv
32. Ebenaceae	Pechi gab[1]	River ebony	66) *Diospyros malabarica*	Fr,Fw, Dy
	Beelati gab[1]	Wood nut	67) *Diospyros blancoi*	Fr, Fw, T, At
33. Elaeocarpaceae	Jalpai[1]	Indian olive	68) *Elaeocarpus robustus*	Fr,A, J, At
34. Euphorbiaceae	Arbarai[1]	Aonla	69) *Phyllanthus acidus*	Fr,Mv
	Latkan[1]	Latkan	70) *Baccaura ramiflora*	Fr, Mv
	Goma[2]	---	71) *Excoecara agallocha*	T, Fw,Iy,H
	Latim[2]	False white teak	72) *Trewia nudiflora*	Fw,P,T
	Urmai[3]	Hurmoi	73) *Sapium indicum*	Bp,Fw
	Patabahar[6]	Croton	74) *Codiaeum variegatum*	F,Hp
	Bhenna[6]	Barbodos nut	75) *Jatropha curcas*	O, Hm,Mv
	Chitki[6]	-----	76) *Phyllanthus chitchensis*	Hm,F
	Amluki[1]	Indiangooseberry	77) *Embelica officinalis*	Fr,Mv, J, A
	Ranghita[6]	Jews slipper	78) *Pedilanthus tithymaloides*	F, Hm
35. Fabaceae	Karanja[2]	Indian buch	79) *Pongamia pinnata*	T,Fw, F, P,Iy,At
	Kata mandar[2]	Coral tree	80) *Erythrin fusca*	T, Fw, F, Mv, Hp
	Palty mandar[2]	Coral tree	81) *Erythrina variegata*	T, Fw, Hp, Mv
	Sissoo[2]	Sissoo	82) *Dalbergia sissoo*	T,F,P,Fw
	Aral dal[6]	Pigeon Pea	83) *Cajanus cajan*	Fr, Mv
	Alkushi[6]	Cow-witch plant	84) *Mucuna pruriens*	Hm,Mv
	Kailla lata[6]	Kalilota	85) *Derris trifoliate*	Mv, F, R
	Ananta kata[6]	-----	86) *Dalbergia spinosa*	Fw
36. Flacourtiaceae	Chaulmoogra[5]	-----	87) *Hydnocarpus kurzii*	Fw, T, Fr, Ir
37. Flagellariaceae	Abeti[6]	Indian *Raton* Lily	88) *Flagellaria indica*	Ir,H, Fr
38. Lamiaceae	Bon tulsi[6]	Wild basil	89) *Ocimum basilicum*	Hm, Mv
	Tulsi[3]	Basil	90) *Ocimum tenuiflorum*	Hm, Mv
39. Lauraceae	Kukurchita[3]	------	91) *Litsea glutinosa*	Hm, Mv
	Tejpata[3]	Bay leaf	92) *Cinnamomum tamala*	S,Mv
	Daruchini[3]	Cinnamon	93) *Chinnamomum verum*	Sp
40. Lecythidaceae	Hijal[5]	Indian oak	94) *Barringtonia acutangula*	Fw,T,Fr
41. Liliaceae	Satamuli[6]	Aparagus	95) *Asparagus racemosus*	Mv
	Gritakumari[6]	Indian aloe	96) *Aleo indica*	Mv, Hm
42. Lythraceae	Mendi[4]	Henna	97) *Lawsonia inermis*	C, Hm,
43. Magnoliaceae	Chapa[4]	Champaka	98) *Michelia champaca*	Fr,P,Fw
44. Malvaceae	Rakta jaba[4]	China rose	99) *Hibiscus rosa-sinensis*	Fr, Rf
	Bhola[5]	Sea hibiscus	100) *Hibiscus tiliaceus*	Fw,T
	Karpash tula[2]	Cotton	101) *Gossypium harbaceun*	Iy, Fw,P
45. Marantaceae	Sitalpati[6]	Murta	102) *Clinogyne dichotoma*	H, M
46. Meliaceae	Mahogony[2]	Mahogany	103) *Swietenia mahagoni*	T, Fe, Fw, Ir, At
	Ghora neem[3]	Barbados lilac	104) *Melia azedarach*	Hm, P, T, Fw
	Neem[3]	Indian mahogany	105) *Toona ciliate*	Hm, P, T, Fw
	Deshi neem[3]	Neem tree	106) *Azadirachta indica*	Hm, P, T, Fw
	Royna[5]	Amoora	107) *Aphamixis polystachya*	P,Fw,H
	Mahogony[2]	Large-leaved Mahogany	108) *Swietenia macrophylla*	T, Fe, Fw, Ir, At
47. Menispermaceae	Paddaguruch/Gula ncha[3]	Gulancha	109) *Tinospora cordifolia*	Hm, Mv

48. Mimosaceae	Akashmoni[2]	Acacia	110) *Acacia auriculiformis*	T, Fe, Fw, P, Wb
	Ipil-Ipil[2]	Ipil-Ipil	111) *Leucaena leucocephala*	Fw,F,Iy
	Jilapi[2]	Jilapi	112) *Pithecellobium dulce*	Fr, T,Fw, P, Mv, H
	Manjium[2]	Mangium	113) *Acacia mangium*	Fr, Fw, P
	Kalo Koroi[2]	Black siris	114) *Albizia odoratissima*	T, Fe, F, P
	Sada Koroi[2]	White Siris	115) *Albizia procera*	T, Fe, F, P
	Raintree[2]	Rain tree	116) *Samanea saman*	Fr, Fw, P, Wb
	Lajjya bati[6]	Sensitible plant	117) *Mimosa pudica*	Mv, Hm
	Chambol[2]	Chapalish	118) *Albizia richardiana*	T, Fw, Fe, P
	Babla[2]	Arabic gum	119) *Acacia nilotica*	T,Fw, Mv, H, Iy
49. Moraceae	Kathal[1]	Jackfruit	120) *Artocarpus heterophyllus*	Fr,Fw, T, F,Fe
	Ban-kathal[5]	----	121) *Artocarpus chaplasha*	A & BF
	Bot[2]	Banyan tree	122) *Ficus benghalensis*	Fw,H,Mv, P, T
	Pakur[2]	Pipal	123) *Ficus religiosa*	Fe, Fw,Mv
	Jag dumur[5]	Fig	124) *Ficus glomerata*	Fr,Fw,Mv
	Dumur[5]	Country fig	125) *Ficus hispida*	Fr, H,Fw
	Harra[5]	Roughbush	126) *Streblus asper*	F, T, B.At
	Deuwa[1]	Monkey jack	127) *Artocarpus lacucha*	Fr,Mv, Hm, Fw
50. Moringaceae	Sajna[1]	Drumstick	128) *Moringa oleifera*	Fr, Vs, Mv
51. Myrtaceae	Deshi Jam[1]	Jamun	129) *Syzygium fruticosum*	Fr,Mv,Fe
	Kalajam[1]	Black berry	130) *Syzygium cumini*	Fr,Mv,Fe
	Jamrul[1]	Wax Jambu	131) *Syzygium samarangense*	Fr,Fw,T
	Peyara[1]	Guava	132) *Psidium guajava*	Fr,Mv,J
	Eucalyptus[2]	Eucalyptus	133) *Eucalyptus camaldulensis*	Fe, P, T
	Golapjam[1]	Rose apple	134) *Syzygm jambos*	Fr,Fw, P
52. Nyctaginaceae	Bagan-bilash[4]	Bougainvillea	135) *Bougainvillea spectabilis*	Fr, Rf
53. Oleaceae	Beli[4]	Arabian Jasmine	136) *Jasminum sambac*	Fr, Rf
54. Pandanaceae	Kewakata[6]	Screwpine	137) *Pandanus foetidus*	Mv,H, A & Bf
55. Poaceae	Basni bash[2]	Bamboo	138) *Bambusa vulgaris*	P, Fw, F, Hp
	Talla bash[2]	Bamboo	139) *Bambusa tulda*	H,Iy, Fw, P
	Akh[6]	Sugar cane	140) *Saccharum officinarum*	J,Hm,F
	Durba[6]	Burmuda grass	141) *Cynodon dactylon*	Mv
	Binnachopa[6]	Vetiver grass	142) *Vetiveria zizanioides*	F, F
	Chau/Chan[6]	Blady grass	143) *Imperata cylindrica*	F
	Tasbidana[6]	Adlay	144) *Coix lacryma-jobi*	Fw, F, Rf
	Nal/Nalkhagra[6]	Tall reed	145) *Phragmites karka*	F,Mv
	Dal gash[6]	-----	146) *Hygroriza aristata*	F
56. Polygonaceae	Biskatali[6]	Pepperwort	147) *Persicaria hydropiper*	Hm, F
57. Pontederiaceae	Kachuripana[6]	Water hyacinth	148) *Eichhornia crassipes*	F, F
58. Portulacaceae	Timeful[6]	Sun plant	149) *Portulaca grandiflora*	Fr, Rf
59. Punicaceae	Dalim[1]	Pomegranate	150) *Punica granatum*	Fr,Mv
60. Rhamnaceae	Boroi[1]	Jujube	151) *Zizyphus mauritiana*	Fr,Fw,Mv,A,P
61. Rosaceae	Golap[6]	Rose	152) *Rosa centifolia*	Fr,C, Rf
	Apple[1]	Apple	153) *Malus sylvestris*	Fr,Mv
62. Rubiaceae	Kelikadam[2]	----	154) *Haldina cordifolia*	T, Fw, Fr
	Gandharaj[4]	Gardenia	155) *Gardenia jasminoides*	Fr, Rf
	Rangon[4]	Rangan	156) *Ixora coccinea*	Fr
	Mewakhathal[5]	---	157) *Morinda angustifolia*	A and BF, Fr
	Kadam[2]	Wild cinchona	158) *Neolamarckia cadamba*	T, Fw, Mv,Iy
63. Rutaceae	Jambura[1]	Pummelo	159) *Citrus maxima*	Fr,Mv, Fw
	Kadbel[1]	Elephant apple	160) *Limonia acidissima*	Fr,Mv, Fw
	Kamla[1]	Orange	161) *Citrus aurantium*	Fr, Mv
	Gora Lebu[1]	Lemon	162) *Citrus lemon*	Fr, Mv
	Lebu/Pati lebu[1]	Common lime	163) *Citrus aurantifolia*	Fr,Mv
	Sarbati Lebu[1]	Sour orange	164) *Citrus aurantium*	Fr, Mv
	Bel[1]	Wood apple	165) *Aegle marmelos*	Fr,Mv,Hm, Fw
64. Sapindaceae	Amjam[6]	-----	166) *Aphania danura*	Fr,Fw
	Litchu[1]	Litchi	167) *Litchi chinensis*	Fr, Fw, F,Fe
65.Sapotaceae	Safeda[1]	Sapota	168) *Manilkara zapota*	Fr, Mv
	Bakul[4]	Indian Medlar	169) *Mimusops elengi*	Fr, Fr

133

66. Scrophulariaceae	Ban-dhane[6]	Sweet broom	170) *Scoparia dulcis*	Hm, Mv
67. Solanaceae	Tita begoon[6]	Black nightshade	171) *Solanum nigrum*	Vg Hm,F
	Dhutra[6]	Angel's Trumpet	172) *Datura metel*	F, Mv
68. Sonneratiaceae	Chaila[5]	Chaila	173) *Sonneratia caseolaris*	Fr, Bf
	Keora[2]	Keora	174) *Sonneratia apetala*	T,F,Fe
69. Sterculiaceae	Ulat-Kambal[3]	Delvil's cotton	175) *Abroma augusta*	Hm, Mv
	Sundari[2]	Sundri	176) *Heritiera fomes*	T,Fw, Fr, P
70. Tiliaceae	Bonpat[6]	Wild jute	177) *Corchorus aestuans*	Fr,F
71. Typhaceae	Hoglapata[6]	Cat tail	178) *Typha domingensis*	H, H, F, M
72. Verbenaceae	Bhait[6]	------	179)*Clerodendrum viscosum*	Hm, F
	Seuli/Sefali[4]	Jasmine	180) *Nyctanthes arbor-tristis*	Fr, C, Rf
	Bain[2]	Baen	181) *Avicennia marima*	T,Fw, F
	Segun[2]	Teak	182) *Tectona grandis*	T,Fe
	Nishinda[3]	Chaste tree	183) *Vitex negundo*	Hm,Hv
	Bamunhatti[6]	Indian tube flower	184) *Clerodendrum indicum*	Fw, A & Bf
73. Vitaceae	Angur[1]	Graps	185) *Vitis vinifera*	Fr,Mv
74. Zingiberaceae	Halud[6]	Turmaric	186) *Curcuma longa*	Sp, Mv
	Ada[6]	Ginger	187) *Zingiber officinale*	Sp, Mv
	Shati[6]	Indian arrawrot	188) *Curcuma zedoaria*	Hm, Fr
	Elachi6[6]	Cardamon	189) *Elettaria cardaomum*	Sp

RP-Relative Prevalence.

-Fruit-yielding species[1], Timber and fuel-yielding species[2], Medicinal and Spice-yielding plant species[3], Ornamental plant species[4], Naturally growing plant species [5], Non-woody plant species [6].

A-Achar, A and BF- Animals & bird food, At- Agricultural tools, Bp-Botanical pesticide, B-Boat, C-Scent/Colour, D-Drink/Dyeing, Fr-Fruit, Flower, Fw-Fuel wood, F-Fodder/Fuel, Fence, Fe-Furniture, Hm-Herbal medicine, H-Handicrafts, Hp-Hedge plants, Hf-Hand fan, Iy-Industry, O-Oil, R-Rope, Rf-Religious festival, Mv- Medicinal value, M-Mat/Molasses, Sp- Spices, T-Timber, P-Pole, J-Jam/Jelly, Vg-Vegetable.

Jjjjjjjjjjjjjjjjjjjjjjjjjjjjjjjjjjjjj

Annexure-4. List of homestead vegetable-yielding species identified and characterized at the study areas of the southwestern coastal region of Bangladesh.

Family	English name	Local name	Scientific name	Cultiv./ Wild	Species in homestead	RP (%) *	Uses
Acrostichaceae	Tiger fern	Oudhachopa	1. *Acrostichum aureum*[1]	Wild	----	----	C,Fr**
Amaranthaceae	Stem amaranth	Danta	2. *Amaranthus lividus*[2]	Cultiv.	66	27.50	Lv,S,C,Mv
	Red amaranth	Lalshak	3. *Amaranthus tricolor*[3]	Cultiv.	89	37.08	Lv,Mv
	-----	Haicha	4. *Alternanthera philoxeroides*[1]	Wild	----	----	LV,FR,MV
Araceae	Taro	Panikachu	5. *Alocasia esculenta var.aquatilis*[2]	Cultiv.	9	3.75	S,C,Mv
	Giant taro	Mankachu	6. *Alocasia macrorrhiza*[1]	Cultiv.	46	19.17	S,C,Sh,Mv
	Yam	Kachu /Boi	7. *Colocasia esculenta*[1]	Wild	----	----	Lv,S,C
	Yam	Biskachu	8. *Colocasia sp.* [1]	Wild	1	0.42	C, Sh,Mv
	Cocoyam	Chinikachu	9. *Xanthosoma atrovirens*[1]	Wild	18	7.50	Lv, C, Sh, Mv
Asteraceae	Water primerose	Helencha	10. *Enhydra fluctuans*[2]	Wild	----	----	Lv,FR,Mv
	Cocklebur	Ghagra	11. *Xanthium strumarium*[3]	Wild	----	----	L,S, C, Sp
Athyriaceae	Table fern	Dhekishak	12. *Diplazium esculentum*[2]	Wild	----	----	Lv,Fr
Basellaceae	Indian spinach	Puishak	13. *Basella rubra*[1]	Cultiv.	81	33.75	Lv,S,C,Fr,M
Caricaceae	Pepey	Pepey	14. *Carica papaya*[1]	Cultiv.	139	57.92	LV,F,C,Sh,Mv
Chenopodiacea e	Spinach	Palong	15. *Beta palonga* [3]	Cultiv.	----	----	Lv.Fr,
	Lambs qarter	Bathua	16. *Chenopodium album*[3]	Wild	----	----	Lv,S,Sp,Mv
Convolvulacea e	Swamp cabbage	Pani kalmi	17. *Ipomoea aquatica*[2]	Wild	4	1.69	Lv,Fr,Mv
	Sweet potato	Mistialu	18. *Ipomoea batatas*[1]	Cultiv.	9	3.75	Lv, Fr, Mv
	Kang Kong	Kalmisak	19. *Ipomoea reptans*[2]	Cultiv.	4	1.67	Lv,FR,Mv
Cucurbitaceae	White Gourd	Chalkumra	20. *Benincasa cerifera*[2]	Cultiv.	25	10.42	Lv,C,B,M
	Cucumber	Sashsa	21. *Cucumis sativus*[2]	Cultiv.	83	34.58	F,C, Mv,M
	Cucumber-short	Khiroi	*Cucumis sativus*	Cultiv.	----	----	F. Mv
	Bottle gourd	Lau/kadu	22. *Lagenaria siceraria*[3]	Cultiv.	146	60.83	LV,S,C,B,Fw, Mv,M
	Ribbed gourd	Jhinga	23. *Luffa acutangula*[2]	Cultiv.	92	38.33	LV,Ss,Fw,H,C, B
	Sponge gourd	Dhundul	24. *Luffa cylindrica* [2]	Cultiv.	28	11.67	C,Sh,Mv
	Bitter gourd	Karala	25. *Momordica charantia*[2]	Cultiv.	56	23.33	Fr,Mv
	Bitter Gourd	Ucchey	*Momordica charantia*	Cultiv.	3	1.25	F,Fr,Mv
	Teasle gourd	Kankrol	26. *Momordica choelis*[2]	Cultiv.	6	2.50	Fr,Mv

	Teasle gourd	Buno Kankrol Wild	*Momordica cochinchinensis*	Wild	----	----	C,Fr,Mv**
	Snake gourd	Rekha	27. *Trichosanthes anguina*[2]	Cultiv.	86	35.83	C,Sh,Mv
Cruciferae	Turnip	Shalgom	28. *Brassica campestris*[3]	Cultiv.	7	2.92	T, Fr, C,Mv
	Cabbage	Bandhakapi	29. *Brassica oleracea*[3] var. *capittata*	Cultiv.	15	6.25	Lv, Fr, Mv
	Cauliflower	Phulkopi	*Brassica oleracea* var. botrydis	Cultiv.	13	5.42	Lv, Fr, Fw
	Radish	Mula	30. *Raphanus sativus*[3]	Cultiv.	41	17.08	Fr, C, Mv
Dioscoreaceae	Winged yam	Mateyalu	31. *Dioscorea alata*[1]	Cultiv.	9	3.75	Mr,C,Sh,Mv
Fabaceae	Sword bean	Mouseem	32. *Canavalia gladiata*[2]	Cultiv.	1	0.42	Fr, C, Sh, Mv
	Country bean	Seem	33. *Lablab purpureus*[3]	Cultiv.	125	52.08	Fr, C, Sh, Mv
	-----	Kalai sak	34. *Lathyrus sativus*[3]	Cultiv.	----	----	Lv, Fr,Mv
	Winged bean	Santarkari	35. *Psophocarpus tetragonolobus*[3]	Cultiv.	1	0.42	Fr, C, Mv
	----	Bakphul	36. *Sesbania grandiflora*[1]	Cultiv.	----	----	F,Fw
	String Bean	Barbati	37. *Vigna sinensis*[2]	Cultiv.	37	15.42	Fr, C, Sh
	Bush bean	Felon	38. *Vigna unguiculata*[3]	Cultiv.			C, Sh, Mv
Gramineae	Sweet corn	Bhutta	39. *Zea mays*[3]	Cultiv.	----	----	F, Fr,
Malvaceae	Ladies Finger	Dherosh	40. *Hibiscus esculentus*[2]	Cultiv.	66	27.50	Fr,C,Sh,Mv
	Roselle	Chukur	41. *Hibiscus sabdariffa*[2]	Cultiv.	----	----	F,C*
Moringaceae	Drumstick	Sajna	42. *Moringa oleifera*[2]	Cultiv.	----	----	F, C,Fr**
Musaceae	Plantain banana	Anajikala	43. *Musa paradisiaca*[1]	Cultiv.	73	30.42	F,C,Fr,Fw,Sh
	Seeded Banana	Aitakala	44. *Musa sylvestris*[1]	Cultiv.	73	30.42	F,C,Fr,Fw,Sh, Mv
Nymphaeaceae	Water lily	Shapla	45. *Nymphoea pubescens*[2]	Wild	----	----	F, S,C,Fw,Mv
Portulaceae	Common purslane	Nonta sak	46. *Portulaca oleracea*[1]	Wild	----	----	Lv,Mv
Solanaceae[3]	Chilli	Khutta marich	47. *Capsicum annum*[1]	Cultiv.	4	1.67	F, C,Sh,Mv ,P
	Chilli	Khara marich	48. *Capsicum frutescens*[1]	Cultiv.	38	15.83	F, C, Sh, P
	Chilli	Bombaimari ch	*Capsicum frutescense* Cv. 'Bombai'	Cultiv.	53	22.08	S,C,Sh,J, P
	Tomato	Tomato	49. *Lycopersicon lycopersicum*[3]	Cultiv.	29	12.08	F, Fr,C,J
	Egg plant	Begoon	50. *Solanum melongena*[1]	Cultiv.	81	33.75	C,Sh,Fr,Mv
	-----	Titbegoon	51. *Solanum torvum*[1]	Wild	----	----	F, C,Sh,Mv**
	Potato	Golalu	52. *Solanum tuberosum*[3]	Cultiv.	2	0.83	T,C,Fr,Sh,Mv
Tiliaceae	Jute leaf	Patshak	53. *Corchorus capsularis*[2]	Cultiv.	10	4.17	Lv,Fr,Mv
Umbelliferae	Coriander	Dhaniapata	54. *Coriandrum sativum*[3]	Cultiv.	4	1.67	Sp,C, Sh, Mv
	Centella	Thankuni	55. *Centella asiatica*[1]	Wild	----	----	Fr,Sh,Mv
	Eryngium	Bilati dhaina	56. *Eryngium foetidum*[1]	Cultiv.	10	4.17	Sp,C, Sh,Mv

136

Year Round[1], Summer[2], and Winter[3] *Relative Preference of the species; ** Very popular to the Rakhain Community; Lv = Leafy vegetable, F= Fruits, Fr. = Fry, Fw = Flower, S = Stem, Sp = Spices, T = Tuber, C = Curry, Mr = Modified root, Mv = Medicinal value, Sh = Smash, B = Bark, M = Morobba (indigenous tasty food prepared with mixing salt and sugar/molasses), J = Jelly, P = Pickle.

Appendix-5: Relative Prevalence (RP) in varying saline zones of southwestern costl region of Bangladesh.

Scientific name	English and local name	Relative prevalence of species in varying saline zones			Average trees	Total	
		Lees saline	Moderately saline	Strongly saline		% of homesteads with the species	RP all farm
1) Timber-yielding species							
Acacia auriculiformis	Akashmoni/Acacia	0.028	0.039	0.035	0.29	0.12	0.03
Acacia nilotica	Babla/Arabic gum	0.001	0.044	0.079	0.31	0.10	0.03
Avicennia marima	Bain/Bean	----	----	0.015	0.08	0.02	0.00
Bambusa tulda	Bash (Basni)/Bambo	0.187	0.331	0.438	1.64	0.24	0.40
Bambusa tulda	Talla Bash /Bambo	0.003	0.027	0.068	0.22	0.12	0.03
Ficus benghalensis	Bot/Banyan tree	0.063	0.028	0.001	0.19	0.12	0.02
Albizia richardiana	Chambol/Chapalish	36.516	21.595	26.130	31.56	0.89	28.01
Alstonia scholaris	Chatian/Devil's tree	0.008	----	----	0.03	0.03	0.00
Eucalyptus camaldulensis	Eucalyptus	0.001	0.004	0.013	0.12	0.04	0.00
Excoecara agallocha	Gewa or Goma	----	0.263	0.031	0.40	0.13	0.05
Leucaena leucocephala	Ipil-Ipil/	0.000	----	0.000	0.01	0.01	0.00
Pithecellobium dulce	Jilapi	0.014	2.423	0.935	2.18	0.37	0.80
Neolamarckia cadamba	Kadam/Wild cinchona	0.041	0.003	0.001	0.14	0.07	0.01
Lannea coromandelica	Kapila or Jiga	0.211	0.154	0.446	1.08	0.25	0.27
Pongamia pinnata	Karanja/Indian buch	0.019	0.595	0.670	1.16	0.29	0.33
Terminalia catappa	Indian almond	0.309	0.545	0.330	0.95	0.41	0.39
Sonneratia apetala	Keora	0.001	0.001	0.005	0.07	0.03	0.00
Albizia odoratissima	Koroi/Black siris	0.243	0.508	0.154	0.99	0.29	0.29
Albizia procera	Sada Koroi	0.198	0.413	0.346	0.98	0.33	0.32
Erythrin fusca	Mandar/Coral tree	0.158	0.021	0.039	0.35	0.18	0.06
Acacia mangium	Mangium	----	----	----	0.00	----	0.00
Swietenia mahagoni	Mehagoni	51.989	15.397	16.350	32.33	0.83	26.67
Oroxylum indicum	Nauasonail	----	----	0.002	0.05	0.00	0.00
Ficus religiosa	Pakur/Pipal	0.002	0.011	0.012	0.10	0.08	0.01
Erythrina variegata	Paltymandar/Coral tree	0.016	0.008	0.008	0.13	0.08	0.01
Trewia nudiflora	Pitali/False white teak	----	----	----	0.00	----	0.00
Samanea saman	Rain tree	16.718	25.686	25.547	24.41	0.93	22.68
Aphamixis polystachya	Royna	----	0.001	0.001	0.03	0.02	0.00
Tectona grandis	Segun/Teak	0.001	0.001	0.000	0.03	0.03	0.00
Dalbergia sissoo	Sissoo	0.009	0.000	0.004	0.09	0.04	0.00
Cassia fistula	Sonail/Indianlaburnum	0.063	0.028	0.001	0.19	0.12	0.02
Heritiera fomes	Sundri	0.000	0.035	0.007	0.20	0.05	0.01
Haldina cordifolia	Kelikadam	----	----	----	0.00	----	0.00
Ceiba pentandra	KatTula/Cotton	0.092	0.692	0.675	1.10	0.40	0.44
Gossypium harbaceun	Karpash Tula /Cotton	0.000	0.009	0.083	0.31	0.06	0.02
Bombax ceiba	Simul tula /Silk cotton	0.058	0.070	0.216	0.76	0.16	0.12
2) Fruit-yielding specie							
Mangifera indica	Aam/Mango	10.620	7.678	14.144	11.92	0.91	10.88
Embelica officinalis	Indian gooseberry	0.038	0.007	0.001	0.13	0.08	0.01
Spondias pinnata	Amra/Hogplum	0.033	0.066	0.374	0.45	0.28	0.12
Vitis vinifera	Angur/Graps	0.001	----	0.000	0.01	0.01	0.00
Malus sylvestris	Apple	0.000		0.000	0.01	0.01	0.00
Phyllanthus acidus	Arbarai/Aonla	----	----	----	0.00	----	0.00
Annona reticulata	Atafol/Bullocks heart	0.003	0.001	0.016	0.10	0.06	0.01
Diospyros blancoi	Beelati gab/Wood nut	0.661	2.145	1.296	3.01	0.44	1.32
Aegle marmelos	Bel/Wood apple	0.009	0.033	0.168	0.29	0.18	0.05
Zizyphus mauritiana	Boroi/Jujube	2.086	1.118	0.207	1.88	0.54	1.02
Dillenia indica	Chalta/Indian dillenia	0.003	0.002	0.072	0.15	0.10	0.01
Averrhoa carambola	Chinese gooseberry	0.003	0.002	0.072	0.15	0.10	0.01
Garcinia cowa	Couphal/Cowea	0.007	0.003	0.009	0.13	0.05	0.01
Punica granatum	Dalim/Pomegranate	0.003	0.038	0.011	0.17	0.09	0.01
Syzygium fruticosum	Deshijam/Jamun	0.963	0.825	0.295	1.33	0.50	0.66
Citrus aurantifolia	Deshilebu/Lemon	0.186	0.093	0.196	0.58	0.28	0.16
Artocarpus lacucha	Monkey jack	0.077	0.118	0.094	0.40	0.24	0.10
Syzygum jambos	Golapjam/Rose apple	0.000	----	0.002	0.02	0.02	0.00
Elaeocarpus robustus	Jalpai/Indian olive	0.001	0.000	0.026	0.07	0.07	0.00
Syzygium wallichii	Jam/Jamun	0.000	0.005	----	0.03	0.03	0.00
Citrus maxima	Jambura/ Pummelo	0.077	0.118	0.094	0.40	0.24	0.10
Syzygium samarangense	Jamrul/Wax jambo	0.108	0.020	0.115	0.32	0.23	0.07
Limonia acidissima	Kadbel/Elephant apple	0.001	0.001	---	0.0	0.0	0.00
Citrus aurantifolia	Kagogilebu/Lemon	0.009	0.013	0.019	0.19	0.07	0.01
Citrus aurantium	Kamla/Orange	0.001	0.000	0.001	0.02	0.02	0.00

Scientific name	Common name						
Artocarpus heterophyllus	Kathal/Jack fruit	1.085	0.951	4.203	3.13	0.61	1.90
Phoenix sylvestris	Khejur/Date palm	10.199	10.766	1.740	8.92	0.78	6.98
Phoenix dactylifera	Khurmakhejur	----	0.001	----	0.01	0.01	0.00
Baccaura ramiflora	Latkan	----	----	----	0.00	----	0.00
Litchi chinensis	Litchu/Litchi	0.015	0.009	0.011	0.13	0.09	0.01
Cocos nucifera	Narikel/Coconut	11.282	8.892	9.901	11.08	0.91	10.06
Diospyros malabarica	Pechi gab/River ebony	0.072	0.684	0.354	1.09	0.30	0.32
Psidium guajava	Peyara/Guava	3.716	3.036	2.750	4.44	0.71	3.16
Manilkara zapota	Safeda/Sapota	0.137	0.004	0.020	0.26	0.15	0.04
Moringa oleifera	Sajna/Drumstick	0.002	0.001	0.000	0.05	0.02	0.00
Citrus aurantium	Sarbati lebu/Lemon	0.000	0.000	0.002	0.03	0.02	0.00
Annona squamosa	Sarifa/Custard apple	0.007	0.031	0.001	0.12	0.08	0.01
Areca catechu	Supari/Betel nut	2.262	2.215	14.142	9.96	0.53	5.31
Borassus flabellifer	Tal/palmyra palm	6.439	5.425	2.738	6.13	0.78	4.78
Tamarindus indica	Tetul/Tamarind	0.797	1.458	0.580	2.05	0.45	0.93
Averrhoa carambola	Chinese gooseberry	0.077	0.118	0.094	0.40	0.24	0.10
3) Medicinal and spice-yielding species							
Calotropis gigantea	Swallow wort	0.020	0.007	0.000	0.10	0.06	0.006
Terminalia arjuna	Arjun/Malabar nut	0.030	0.002	0.021	0.16	0.10	0.015
Justicia adhatoda	Basak	0.003	0.001	0.001	0.04	0.03	0.001
Terminalia bellerica	Bohera/Belleric myrobalan	0.001	0.000	0.000	0.02	0.02	0.000
Chinnamomum verum	Daruchini/ Cinnamon	0.001	0.000	0.001	0.03	0.03	0.001
Azadirachta indica	Deshi neem/Neem	1.478	2.040	0.065	1.91	0.49	0.938
Elettaria cardaomum	Elachi/Cardamon	----	----	----	0.00	----	0.00
Melia azedarach	Ghora neem/Bread tree	0.001	0.004	0.000	0.03	0.03	0.001
Terminalia chebula	Hartaki/ Black myrobalan	0.001	0.000	0.000	0.02	0.02	0.000
Litsea glutinosa	Lot-pipal	0.000	----	0.025	0.07	0.05	0.003
Vitex negundo	Nishinda/Chase tree	0.000	0.001	0.001	0.03	0.03	0.001
Tinospora cordifolia	Paddaguruch	0.000	0.000	----	0.01	0.01	0.000
Cinnamomum tamala	Tejpata/Bay leaf	0.008	0.000	0.010	0.07	0.06	0.004
Ocimum tenuiflorum	Tulsi/Basil	0.001	0.004	0.000	0.03	0.03	0.001
Abroma augusta	Delvil's cotton	0.000	----	0.002	0.02	0.02	0.000
Sapium indicum	Urmai/Hurmoi	0.013	0.001	0.000	0.08	0.03	0.003
4) Ornamental plant species							
Mimusops elengi	Bakul/Indian medlar	0.0025	0.0063	0.0013	0.07	0.05	0.0031
Jasminum sambac	Beli/Arabian jasmine	0.0039	0.0047	0.0019	0.06	0.05	0.0034
Bougainvillea spectabilis	Bougainvillea	0.0014	0.0100	0.0014	0.06	0.06	0.0034
Michelia champaca	Chapa/Champaka	0.0044	0.0006	0.0002	0.04	0.03	0.0012
Polyalthia longifolia	Debdaru/Most tree		0.0005		0.01	0.00	0.0001
Gardenia jasminoides	Gandharaj/Gardenia	0.0056	0.0531	0.0047	0.13	0.12	0.0156
Hibiscus rosa-sinensis	Jaba/China rose	0.0183	0.0506	0.0600	0.23	0.18	0.0411
Casuarina littorea	Jhau/Seef wood	--	0.0014	0.0002	0.02	0.02	0.0003
Neium indicum	Karabi/Oleander	0.0002	----	0.0006	0.01	0.01	0.0002
Thevetia peruviana	Karali/ Lucky nut	0.0016	0.0063	----	0.05	0.03	0.0016
Delonix regia	Krisnachura/ Gulmohur	0.0127	0.0077	----	0.07	0.07	0.0044
Lawsonia inermis	Mendi/Henna	0.0127	0.0056	0.0306	0.12	0.12	0.0146
Catharanthus roseus	Nayantara/Periwinkle	----	----	----	0.00	----	0.00
Codiaeum variegatum	Patabahar /Croton	----	----	----	----	----	----
Polianthes tuberosa	Rajanigandha/ Tube rose	----	----	----	----		----
Ixora coccinea	Rangan	0.0006	0.0002	----	0.01	0.01	0.0002
Nyctanthes arbor-tristis	Seuli/Jasmine	0.0002	0.0025	0.0025	0.04	0.04	0.0014
5) Naturally growing species							
Cordia dichotoma	Bahal	----	----	----	0.00	----	0.00
Hibiscus tiliaceus	Balai-gach/ Sea hibiscus	0.075	0.008	0.000	0.17	0.09	0.016
Hydnocarpus kurzii	Bal-gach	0.040	0.017	0.038	0.23	0.13	0.031
Spondias dulcis	Bunoamra/Wild hogpalm	0.002	0.002	0.003	0.05	0.04	0.002
Artocarpus chaplasha	Ban-kathal	0.020	0.000	----	0.07	0.04	0.003
Sonneratia caseolaris	Chaila	---	0.007	0.004	0.07	0.04	0.003
Ficus hispida	Dumur/Country fig	0.025	0.008	0.047	0.23	0.11	0.025
Calophyllum inophyllum	Gaitta/ Borneo mahogany	0.001	0.001	----	0.03	0.02	0.000
Lannea coromandelica	Gigni	----	0.001	0.009	0.06	0.03	0.002
Streblus asper	Harra/Roughbush	0.152	0.005	0.425	0.61	0.23	0.138
Barringtonia acutangula	Hijal/ Indianoak	0.011	0.001	0.069	0.19	0.09	0.018
Ficus glomerata	Jagdumur/Fig	0.000	0.031	0.014	0.15	0.07	0.011
Morinda angustifolia	Mewakhathal	----	0.00	----	0.00	----	0.00
Aphamixis polystachya	Royna	0.001	0.006	0.020	0.10	0.07	0.007

139

Appendix- 6: Type of tree species planted during the last 10 years at different farm category of the study areas of the southwestern coastal zone of Bangladesh.

Local name	Plantation/homestead during last 10 years				Group Total	Rank
	Landless	Small	Medium	Large		
Timber-yielding species						
Mahogony	10.73	18.78	36.60	32.17	23.30	1
Chambol	9.21	18.13	35.38	31.53	22.33	2
Raintree	9.06	18.72	21.63	19.83	17.65	3
Bash (Basni)	0.04	1.02	0.07	0.17	0.48	4
Koroi (Kalo)	0.31	0.39	0.37	0.67	0.40	5
Tejpata	0.00	0.49	0.00	1.03	0.34	6
Telikadam	0.10	0.18	0.48	0.33	0.26	7
Akashmoni	0.00	0.14	0.17	0.43	0.15	8
Tula	0.04	0.15	0.25	0.17	0.15	9
Deshi Neem	0.00	0.16	0.03	0.63	0.15	10
Sissoo	0.06	0.27	0.00	0.00	0.13	11
Pahari Neem	0.00	0.20	0.00	0.17	0.10	12
Mandar (Kata)	0.06	0.02	0.17	0.17	0.08	13
Katbadam	0.08	0.10	0.02	0.03	0.07	14
Sonail	0.00	0.03	0.18	0.10	0.07	15
Eucalyptus	0.00	0.09	0.00	0.10	0.05	16
Koroi (Sada)	0.04	0.05	0.10	0.00	0.05	17
Arjun	0.00	0.02	0.12	0.00	0.04	18
Elachi	0.00	0.00	0.00	0.33	0.04	19
Kadam	0.10	0.00	0.05	0.00	0.03	20
Bash (Talla)	0.00	0.02	0.00	0.00	0.01	21
Bot	0.00	0.00	0.00	0.03	0.00	22
Chatian	0.00	0.00	0.00	0.00	0.00	23
Minjiri	0.00	0.00	0.00	0.00	0.00	24
Hartaki	0.00	0.00	0.00	0.03	0.00	25
Ulat-Kambal	0.00	0.00	0.00	0.00	0.00	26
Daruchini	0.00	0.00	0.00	0.00	0.00	27
Baganbilash	0.00	0.00	0.00	0.00	0.00	28
Chapa(kathali)	0.00	0.00	0.00	0.00	0.00	29
Jaba	0.00	0.00	0.00	0.00	0.00	30
Seuli	0.00	0.00	0.00	0.00	0.00	31
Gandharaj	0.00	0.00	0.00	0.00	0.00	32
Golap	0.00	0.00	0.00	0.00	0.00	33
Krisnachura	0.00	0.00	0.00	0.00	0.00	34
Karabi	0.00	0.00	0.00	0.00	0.00	35
Sub total	29.83	58.96	95.62	87.92	65.88	----
Fruit-yielding species						
Aam	3.67	6.91	5.18	10.13		1
Narikel	3.29	4.27	5.52	3.03	4.23	2
Supari	1.42	3.85	6.63	3.47	4.01	3
Peyara	2.08	2.92	2.10	4.13	2.70	4
Kathal	1.46	1.70	1.85	5.23	2.13	5
Tal	0.46	0.78	2.10	1.53	1.14	6
Khejur	0.67	0.99	1.40	1.50	1.09	7
Lebu	0.31	0.91	0.55	0.87	0.70	8

Jambura	0.56	0.45	0.70	1.43	0.66	9
Boroi	0.46	0.38	0.95	0.43	0.55	10
Jam	0.15	0.59	0.40	0.67	0.46	11
Tetul	0.27	0.45	0.62	0.43	0.45	12
Amra	0.19	0.13	0.32	0.47	0.23	13
Jmrul	0.13	0.24	0.13	0.33	0.20	14
Litchu	0.04	0.20	0.22	0.40	0.20	15
Kamranga	0.08	0.16	0.23	0.20	0.17	16
Safeda	0.06	0.20	0.18	0.07	0.15	17
Dalim	0.00	0.18	0.27	0.00	0.14	18
Jalpai	0.00	0.06	0.23	0.23	0.11	19
Amluki	0.08	0.13	0.03	0.07	0.09	20
Bel	0.06	0.07	0.12	0.13	0.09	21
Others	0.00	0.17	0.03	0.00	0.08	22
Sarifa	0.04	0.08	0.07	0.03	0.06	23
Deuwa	0.04	0.05	0.03	0.07	0.05	24
Atafol	0.06	0.03	0.03	0.07	0.04	25
Chalta	0.04	0.02	0.03	0.00	0.03	26
Kadbel	0.00	0.02	0.03	0.00	0.02	27
Angur	0.00	0.00	0.00	0.00	0.00	28
Sub total	15.63	25.92	29.97	34.93	26.00	-----
Grand total	45.46	84.88	125.59	122.85	91.88	64

Appendix-7: Felling trend of trees in the homesteads during the last 10 years at different farm category in the study areas of the coastal zone of Bangladesh.

Local name	Trees felled/homestead during last 10 years			
	Landless	Small	Medium	Large
Timber-yielding tree species				
Raintree	11(2.67)	24 (4.98)	16 (4.83)	15 (4.30)
Chambol	15 (1.25)	20 (1.19)	7 (1.25)	10 (2.13)
Mahogony	10 (0.96)	20 (1.27)	6 (0.80)	15 (2.83)
Babla	5 (0.19)	4 (0.10)	7 (0.23)	10 (0.47)
Telikadam	7 (0.44)	5 (0.22)	5 (0.33)	4 (0.23)
Jilapi.	10 (0.77)	10 (0.70)	10 (0.60)	5 (0.60)
Koroi	2 (0.08)	6 (0.47)	7 (0.480	8 (0.87)
Katbadam	1 (0.04)	2 (0.05)	3 (0.17)	2 (0.17)
Karanja	2 (0.08)	0 (0.00)	3 (0.10)	2 (0.13)
Keora	7 (0.15)	1 (0.01)	0 (0.00)	0 (0.00)
Mandar	10 (0.35)	2 (0.04)	1 (0.02)	0 (0.00)
Gewa	4 (0.13)	0 (0.00)	2 (0.07)	2 (0.07)
Chatian	2 (0.08)	1 (0.01)	0 (0.00)	0 (0.00)
Eucalyptus	5 (0.10)	0 (0.00)	0 (0.00)	0 (0.00)
Akashmoni	1 (0.02)	5 (0.08)	0 (0.00)	0 (0.00)
Tula	2 (0.10)	4 (0.06)	1 (0.07)	0 (0.00)
Pakur	1 (0.02)	1 (0.01)	0 (0.00)	0 (0.00)
Bot	1 (0.02)	0 (0.00)	0 (0.00)	0 (0.00)
Royna	0 (0.00)	0 (0.00)	0 (0.00)	0 (0.00)
Sissoo	1 (0.02)	3 (0.03)	0 (0.00)	0 (0.00)
Sonail	1 (0.02)	0 (0.00)	0 (0.00)	1 (0.07)
Sundari	0 (0.00)	6 (0.06)	0 (0.00)	0 (0.00)
Arjun	0 (0.00)	0 (0.00)	0 (0.00)	0 (0.00)
Neem	4 (0.10)	1 (0.03)	5 (0.12)	0 (0.00)
Fruit-yielding tree species				
Aam	10 (0.58)	10 (0.490	5 (0.23)	10 (0.93)
Jam	4 (0.08)	8 (0.10)	1 (0.02)	1 (0.10)
Tal	2 (0.23)	11 (0.57)	7 (0.77)	6 (0.67)
Tetul	1 (0.04)	7 (0.18)	3 (0.10)	3 (0.37)
Khejur	5 (0.33)	20 (0.74)	13 (0.92)	5 (0.57)
Narikel	2 (0.08)	8 (0.29)	6 (0.28)	7 (0.63)
Kathal	0 (0.00)	1 (0.01)	0 (0.00)	1 (0.07)
Boroi/Kul	1 (0.02)	2 (0.12)	2 (0.13)	4 (0.17)
Bilati gab	0 (0.00)	1 (0.01)	2 (0.03)	2 (0.07)
Pechi gab	0 (0.00)	5 (0.11)	0 (0.00)	0 (0.00)
Supari	5 (0.13)	10 (0.20)	2 (0.03)	9 (0.47)
Others	0 (0.00)	4 (0.07)	0 (0.00)	1 (0.03)

Appendix-8. Homestead plant species deterioration compared to the past in the study areas of the southwestern coastal region of Bangladesh.

English name/ Bengali name	Scientific name	Respondents opinion regarding species changing over time			
		Species decreasing 2008		Species decreasing 1988	
		Rank	RR*	Rank	RR*
Palmyra palm	*Borassus flabellifer*	87	36	120	50
Date palm	*Phoenix sylvestris*	80	33	131	55
Sundri	*Heritiera fomes*	75	31	136	57
River ebony	*Diospyros blancoi*	71	29	119	50
Cowa	*Garcinia cowa*	69	27	99	41
Betel nut	*Areca catechu*	69	29	112	47
Jamun	*Syzygium fruticosum*	68	28	116	48
Devils tree	*Alstonia scholaris*	68	28	90	37
Indian dillenia	*Dillenia indica*	65	20	78	32
Indian buch	*Pongamia pinnata*	62	26	109	45
Indian almond	*Terminalia catappa*	62	26	77	32
Telikadam	*Adina cordifolia*	61	26	79	33
Mango	*Mangifera indica*	60	25	120	50
Been/Bain	*Avicennia officinalis*	60	25	79	33
Monkey jack	*Artocarpus lacucha*	60	25	72	30
Indian laburnum	*Cassia fistula*	60	21	74	31
Seeded banana	*Musa sylvestris*	60	29	102	42
Indian almond	*Neolamarckia cadamba*	56	23	90	38
Karoj	*Albizia proceera*	55	23	76	32
Jalpai	*Elaeocarpus robustus*	53	22	85	35
Coconut	*Cocos nucifera*	53	22	101	42
Tamarind	*Tamarindus indica*	50	21	97	40
Custard apple	*Annona reticulata*	50	29	78	33
Royna	*Apharamixis polystachya*	50	21	60	25
Water lily	*Nymphaea nauchalli, N. pubeseence*	48	20	96	40
Wood apple	*Aegle marmelos*	48	12	55	22
Keora	*Sonneratia apetala*	45	19	73	30
Malabur nut	*Terminalia arjuna*	45	19	61	25
Pome granate	*Punica granatum*	42	17	50	21
Gooseberry	*Averrhoa carambola*	42	17	60	25
Bamboo	*Bambusa vulgaris*	41	17	98	41
Neem	*Azadirachta indica*	40	17	97	40
Fig	*Ficus hispida*	40	21	50	21
Custard apple	*Annona squamosa*	36	15	50	21
Nisinda	*Vitex negundo*	32	13	62	26
Jujube	*Zizyphus mauritiana*	30	16	50	21
Guava	*Psidium guajava*	30	13	39	16
Indian oak	*Barringtonia acutangula*	29	12	40	17
Basak	*Adhatoda vasica*	28	12	51	21
Jilapi	*Pithecellobium dulce*	28	12	42	17
Cane	*Calamus rotung*	25	11	60	26
Coral tree	*Erythrina veriegata*	25	11	66	17
Biralkata	*Mucuna pruriens*	22	9	34	14

Basil	*Ocimum sanctum*	22	9	60	25
Lucky nut	*Thevetia peruviana*	19	07	25	10
Pummelo	*Citrus maxima*	18	8	28	12
Cotton	*Ceiba pentandra*	17	7	77	32
Screwpine	*Pandanus foetidus*	15	6	10	04
Sarnalat	*Cuscuta reflexa*	15	6	----	----
Pigeon Pea	*Cajanus cajan*	15	6	11	06
Paddagurs	*Tinospora cordifolia*	15	6	-----	----
Lat-pipal	*Litsea glutinosa*	14	6	02	.83
Plantan banana	*Musa paradisiacal*	11	5	17	07
Hurmai	*Sapium indicum*	11	6	02	0.83
Delvil's cotton	*Abroma augusta*	10	4	05	02
Winged yam	*Dioscorea alata*	10	4	10	04
Pineapple	*Ananas comosus*	10	4	----	----
Swallow nut	*Calotropis gigantea*	10	4	11	05
Country fig	*Ficus glomerata*	09	4	22	09
Drumstick	*Moringa olerfera*	06	3	02	0.83
Pitali	*Trewia polycarpa*	06	3	06	03
Banyan tree	*Ficus benghalensis*	05	3	10	04

* RR-Relative rank of the respondent opinion.